Antje Maria Moffat

Purely Data-Driven Characterization of Ecosystem Responses

Antje Maria Moffat

Purely Data-Driven Characterization of Ecosystem Responses

A New Methodology

Südwestdeutscher Verlag für Hochschulschriften

Impressum / Imprint

Bibliografische Information der Deutschen Nationalbibliothek: Die Deutsche Nationalbibliothek verzeichnet diese Publikation in der Deutschen Nationalbibliografie; detaillierte bibliografische Daten sind im Internet über http://dnb.d-nb.de abrufbar.
Alle in diesem Buch genannten Marken und Produktnamen unterliegen warenzeichen-, marken- oder patentrechtlichem Schutz bzw. sind Warenzeichen oder eingetragene Warenzeichen der jeweiligen Inhaber. Die Wiedergabe von Marken, Produktnamen, Gebrauchsnamen, Handelsnamen, Warenbezeichnungen u.s.w. in diesem Werk berechtigt auch ohne besondere Kennzeichnung nicht zu der Annahme, dass solche Namen im Sinne der Warenzeichen- und Markenschutzgesetzgebung als frei zu betrachten wären und daher von jedermann benutzt werden dürften.

Bibliographic information published by the Deutsche Nationalbibliothek: The Deutsche Nationalbibliothek lists this publication in the Deutsche Nationalbibliografie; detailed bibliographic data are available in the Internet at http://dnb.d-nb.de.
Any brand names and product names mentioned in this book are subject to trademark, brand or patent protection and are trademarks or registered trademarks of their respective holders. The use of brand names, product names, common names, trade names, product descriptions etc. even without a particular marking in this works is in no way to be construed to mean that such names may be regarded as unrestricted in respect of trademark and brand protection legislation and could thus be used by anyone.

Coverbild / Cover image: www.ingimage.com

Verlag / Publisher:
Südwestdeutscher Verlag für Hochschulschriften
ist ein Imprint der / is a trademark of
AV Akademikerverlag GmbH & Co. KG
Heinrich-Böcking-Str. 6-8, 66121 Saarbrücken, Deutschland / Germany
Email: info@svh-verlag.de

Herstellung: siehe letzte Seite /
Printed at: see last page
ISBN: 978-3-8381-3444-4

Zugl. / Approved by: Jena, FSU, Dissertation, 2012

Copyright © 2013 AV Akademikerverlag GmbH & Co. KG
Alle Rechte vorbehalten. / All rights reserved. Saarbrücken 2013

Danksagungen

Hiermit möchte ich mich ganz herzlich bei allen bedanken, die meine Promotion begleitet haben.

Mein besonderer Dank gilt meinen Betreuern Martin Heimann und Clemens Beckstein für die Freiheit und Unterstützung, genau dieses mich nach wie vor fesselnde Thema zu vertiefen. Galina Churkina danke ich für ihre beratende Unterstützung. Da ich über längere Zeiträume von zu Hause aus gearbeitet habe, war der gute Austausch mit meinen Kollegen am MPI für Biogeochemie und der Arbeitsgruppe Künstliche Intelligenz an der FSU Jena besonders wichtig.

Für das Korrekturlesen dieses Manuskriptes bedanke ich mich bei Myroslava Khomik und Sönke Zaehle, bei Clemens Beckstein für seine unermüdlichen Anmerkungen zur inhaltlichen Geschlossenheit und bei Anton Moffat für das muttersprachliche Aufpolieren.

Das Fertigstellen dieser Arbeit war nur möglich, weil ich daran immer wieder völlig zurückgezogen arbeiten konnte. Ich danke meinem Onkel Gerd Wetzig in Dresden für den 'Elfenbeinkeller' und meinen reisefreudigen Freunden Antje und Markus Sticker für ihre 'Elfenbeinwohnung'. Meiner Mutter danke ich für das Enkeleinhüten in der Abgabephase und ihre fortwährende Unterstützung.

Meinem Vater möchte ich an dieser Stelle für seine vorausschauende Idee danken, mein technisches Interesse zu fördern, indem er mir bereits 1987 einen PC(AT) schenkte.

Ein ganz liebevoller und herzlicher Dank gilt meinem Mann Anton und meinen beiden Kindern Anna und Leo, die die Herausforderung mitgetragen haben, Promotion und Familie unter einen Hut zu bringen.

Original title of the dissertation:
"A new methodology to interpret high resolution measurements of net carbon fluxes between terrestrial ecosystems and the atmosphere"

Also available for download in PDF format:
http://www.db-thueringen.de/servlets/DocumentServlet?id=20321

Contents

List of Figures v
List of Acronyms and Variables ix

1 Introduction   **1**
   1.1 Background: The global carbon cycle 1
   1.2 Data domain 3
       1.2.1 Terrestrial carbon flux measurements 3
       1.2.2 Description of the Hainich forest tower site 5
       1.2.3 Properties of the eddy flux measurements 6
   1.3 Methodological concept 8
       1.3.1 Data problem class 8
       1.3.2 Inductive modeling approach 8
       1.3.3 Characterization with artificial neural networks 9
       1.3.4 Basic principle 11
   1.4 Outline of the thesis 12

I Methodology   **15**

2 Modeling framework   **17**
   2.1 Feedforward artificial neural networks 18
   2.2 Backpropagation algorithm 18
       2.2.1 Network topology 18
       2.2.2 The error function 20
       2.2.3 Online and epoch mode 20
       2.2.4 Momentum term 21
       2.2.5 Regularization term 21
   2.3 Partial derivative propagation 22

2.4	Node properties	24
	2.4.1 Activation function	24
	2.4.2 Scaling of the training pattern	25
	2.4.3 Goodness factor	26
2.5	Training process	26

3 Data analysis toolbox 29
3.1 Mapping performance measures 29
3.2 Driver relevance 31
    3.2.1 Added performance 31
    3.2.2 Greedy search algorithm 32
3.3 Network function 32
3.4 Numerical partial derivatives 33
    3.4.1 Dynamic range normalization 33
    3.4.2 Mean derivatives 35

4 Phases of the methodology 37
4.1 Specifying the response query 38
4.2 Generating driver candidates 39
4.3 Benchmarking with all drivers 40
4.4 Identifying the driver hierarchy 40
4.5 Extracting the functional relationships 41
4.6 Analyzing the sensitivities 42
4.7 Discussing the induced hypotheses 43

II Areas of application 45

5 Characterizing ecosystem responses 47
5.1 Specifying the response query: *Daytime NEP response* 47
5.2 Generating driver candidates 48
5.3 Benchmarking with all drivers 49
5.4 Identifying the driver hierarchy 51
5.5 Extracting the functional relationships 56
    5.5.1 One-dimensional response to light 56
    5.5.2 Multi-dimensional response functions 58
5.6 Analyzing the sensitivities 66

CONTENTS

    5.7 Discussing the induced hypotheses 67

6 Testing specific hypotheses   71
    6.1 Specifying the response query: *Net effect of diffuse radiation* . . 71
    6.2 Generating driver candidates 72
    6.3 Benchmarking with all drivers 72
    6.4 Extracting the functional relationships 73
    6.5 Discussing the induced hypotheses 76

7 Assessing competing semi-empirical equations   77
    7.1 Specifying the response query: *Light response curve* 78
    7.2 Benchmarking with all drivers 80
    7.3 Extracting the functional relationships 81
    7.4 Discussing the induced hypotheses 83
        7.4.1 Equation 1: Linear function with upper limit 84
        7.4.2 Equation 2: Rectangular hyperbola 85
        7.4.3 Equation 3: Modified rectangular hyperbola 86
        7.4.4 Equation 4: Non-rectangular hyperbola 87
        7.4.5 Equation 5: Smith sigmoid 88
        7.4.6 Equation 6: Logistic sigmoid 89
        7.4.7 Equation 7: Exponential saturation 90

8 Evaluating ecosystem models   93
    8.1 Specifying the response query: *Daytime NEP modeled by two TBMs* 93
        8.1.1 Model 1: ORCHIDEE 94
        8.1.2 Model 2: BETHY 94
    8.2 Generating driver candidates 95
    8.3 Benchmarking with all drivers 97
    8.4 Identifying the driver hierarchy 99
    8.5 Extracting the functional relationships 101
    8.6 Discussing the induced hypotheses 104

9 Interpolating missing data   107
    9.1 Specifying the response query: *Gap-filling of missing NEP* . . . 107
    9.2 Generating driver candidates 109
    9.3 Benchmarking with all drivers 111

9.4 Discussing the induced hypotheses 111

III Reliability and conclusions   115

10 Reliability of the modeling results   117
10.1 Cross-validation 117
10.2 Example for regularization 119
10.3 Impact of input uncertainty 121
10.4 Overall assessment 122

11 Conclusions and outlook   125

A Technical implementation   127
A.1 Object-oriented network implementation 127
A.2 Program structure 129
A.3 Program flow 130
A.4 Example of a pattern generation script 131
A.5 Example of a typical specification file 132

Bibliography   135

List of Figures

1.1	Overview of the global carbon fluxes	2
1.2	Sources and sinks of atmospheric CO_2	3
1.3	Sketch of flux tower measurements	4
1.4	Photo of the flux tower at Hainich forest	5
1.5	Conceptual flow of the two modeling approaches	9
1.6	Mathematical model types	10
1.7	Schematic of an ANN model trained on eddy flux measurements	10
1.8	Observation and model simulation streams	11
2.1	Example of a feedforward ANN	19
2.2	Diagram for the feedforward and backpropagation step	22
2.3	Diagram for the error and network output function composition	23
2.4	Graph of the logistic sigmoid activation function	24
2.5	Modeled cubic function with different output node and scaling setups	25
2.6	Typical example for an ANN training progression	28
3.1	Sketch to illustrate the effect of normalization	34
4.1	The seven phases of the methodology	37
5.1	Benchmarking of the daytime NEP response	50
5.2	Standard deviation of the model residuals	51
5.3	Primary driver performance	51
5.4	Scatterplot matrix of the radiative drivers	52
5.5	Secondary driver performance	54
5.6	Tertiary driver performance	55
5.7	Structure of an ANN modeling $NEP(PPFD)$	56
5.8	Modeled response and derivatives of $NEP(PPFD)$	57
5.9	Scatterplot matrix of the meteorological drivers	59

LIST OF FIGURES

5.10	Structure of an ANN modeling $NEP(PPFD_{dir}, PPFD_{dif})$	60
5.11	Modeled response and derivatives of $NEP(PPFD_{dir}, PPFD_{dif})$.	61
5.12	3D-plot of the modeled response $NEP(PPFD_{dir}, PPFD_{dif})$ ...	62
5.13	Structure of an ANN modeling $NEP(PPFD_{dir}, PPFD_{dif}, VPD)$.	63
5.14	Modeled response and derivatives of $NEP(PPFD_{dir}, PPFD_{dif}, VPD)$	64
5.15	Modeled response and derivatives of $NEP(PPFD_{dir}, PPFD_{dif}, VPD)$	65
5.16	Sensitivity of NEP to $PPFD_{dir}$ and $PPFD_{dif}$ versus total $PPFD$	66
5.17	Monthly sensitivity of NEP to $PPFD_{dir}$, $PPFD_{dif}$, and VPD .	67
6.1	Scatterplot matrix of R_{pot}, f_{dif}, and VPD	72
6.2	Network structure and scatterplot of $NEP(R_{pot}, f_{dif}, VPD)$ ...	73
6.3	Modeled response and derivatives of $NEP(R_{pot}, f_{dif}, VPD)$ ...	74
6.4	Modeled response and derivatives of $NEP(R_{pot}, f_{dif}, VPD)$ ...	75
6.5	3D-plot of $NEP(R_{pot}, f_{dif})$ with fixed VPD	75
7.1	Sketch of a typical light response curve $NEP(PPFD)$	78
7.2	Standard deviation of the ANN model residuals of $NEP(PPFD)$	81
7.3	The ANN light response curve and its numerical derivatives ..	82
7.4	ANN and semi-empirical light response curves with their derivatives	86
7.5	Non-rectangular hyperbola for various values of the curvature parameter	88
7.6	Graph of the logistic sigmoid function	89
8.1	Scatterplot of simulated vs. measured NEP for ORCHIDEE and BETHY	95
8.2	Benchmarking of DATA, ORCHIDEE, and BETHY	98
8.3	Primary and secondary driver performance	100
8.4	Modeled response and derivatives of $NEP(PPFD, VPD)$	102
8.5	Modeled response and derivatives of $NEP(PPFD, T_a)$	103
9.1	Effect of input variables describing the yearly ecosystem state .	110
9.2	R^2, $SDev$, and $Bias$ for the ten artificial gap permutations ...	112
9.3	Course of the half-hourly NEP over time	113
9.4	Comparison of $RMSE$ performance with other gap-filling techniques	114
10.1	Cross-validation with a different year and a different site	118
10.2	Cross-validation with and without weight regularization	120
10.3	R^2 performance of the ANN models trained with artificial noise	120

LIST OF FIGURES

10.4	Modeled response and derivatives with correlated noise added .	122
A.1	Inheritence scheme of the backpropagation node classes	128
A.2	Sketch of the program structure	129
A.3	Program flow schematic	130
A.4	Pattern generation script	133
A.5	Pattern processing routine specifications	134

LIST OF FIGURES

List of Acronyms and Variables

Acronyms
    CO_2 Carbon Dioxide
    ANN   Artificial Neural Network
    BP   Backpropagation
    NLR   Non-linear Regression
    TBM   Terrestrial Biosphere Model

Performance Measures
    R^2 Coefficient of Determination
    MAE Mean Absolute Error
    $SDev$ Standard Deviation
    SSE Sum of Squared Errors
    $RSME$ Root Mean Square Error
    $Bias$ Bias Error

Flux Variables (Unit: 1.0 μmol CO_2 m^{-2} s^{-1} = 1.04 g C m^{-2} d^{-1})
    NEE Net Ecosystem Exchange
    NEP Net Ecosystem Productivity ($= -NEE$)
    GPP Gross Primary Production
    ER Ecosystem Respiration

LIST OF ACRONYMS AND VARIABLES

Radiative Variables
- Rg (Total) Global Radiation (W m^{-2})
- Rd Diffuse Global Radiation (W m^{-2})
- Rn Net Radiation (W m^{-2})
- Rr Reflected Radiation (W m^{-2})
- $PPFD$ (Total) Photosynthetic Photon Flux Density (μmol photon m^{-2} s^{-1})
- $PPFD_{dir}$ Direct $PPFD$ (μmol photon m^{-2} s^{-1})
- $PPFD_{dif}$ Diffuse $PPFD$ (μmol photon m^{-2} s^{-1})

Meteorological Variables
- Rh Relative Humidity (%)
- VPD Vapor Pressure Deficit (hPa)
- SWC Soil Water Content (%)
- Ta Air Temperature (°C)
- T_m Mean Daytime Air Temperature (°C)
- Tc Radiative Canopy Temperature (°C)
- Ts_1, Ts_2 Soil Temperature at 5 and 30 cm Depths (°C)
- Gs Soil Heat Flux (W m^{-2})
- $Precip$ Precipitation (mm)
- WD Wind Direction (°)
- WS Wind Speed (m s^{-1})
- $ustar$ Friction Velocity (m s^{-1})
- ZL Atmospheric Stability Parameter

Theoretical Variables
- R_{pot} Potential Radiation (Insolation) at the Top of the Atmosphere (W m^{-2})
- f_{dif} Diffuse Fraction (%)
- $Fuzzy$ Fuzzy Variable for the Time of Day

LIST OF ACRONYMS AND VARIABLES

Physiological Properties of the Light Response Curve
    NEP_{sat} Saturated NEP (μmol CO_2 m^{-2} s^{-1})
    GPP_{opt} Optimum GPP (μmol CO_2 m^{-2} s^{-1})
    ER_{dayt} Daytime ER (μmol CO_2 m^{-2} s^{-1})
    α Initial Apparent Quantum Yield
        (μmol CO_2 m^{-2} s^{-1}/μmol photon m^{-2} s^{-1})
    $PPFD_{com}$ Light Compensation Point (μmol photon m^{-2} s^{-1})
    $PPFD_{sat}$ Light Saturation Point (μmol photon m^{-2} s^{-1})
    $PPFD_{turn}$ Light Turning Point (μmol photon m^{-2} s^{-1})

Chapter 1

Introduction

To understand more about the global carbon cycle, high resolution measurements of the carbon exchange (fluxes) between terrestrial ecosystems and the atmosphere are taken at many sites in the world. Due to the complexity of the system measured and to limitations of the measurement technique itself, the obtained ecosystem datasets are highly complex, noisy, and even fragmented. The underlying causalities cannot be obtained just by visual evaluation of the measurements, but require additional modeling.

The methodology developed here attempts to provide a systematic body of inductive modeling procedures to extract these causalities with as few prior assumptions as possible. The underlying ecophysiological relationships, such as the hierarchy of the climatic controls or their multivariate dependencies, are extracted *directly* from the ecosystem dataset. This new access to the data enables a wide range of ecophysiological applications, as will be demonstrated on carbon flux measurements from the Hainich forest site.

This introductory chapter provides some background on the global carbon cycle (Section 1.1), a description of the data domain (Section 1.2), the basic concept of the methodology (Section 1.3), and an outline of the thesis (Section 1.4).

1.1 Background: The global carbon cycle

Because of its large abundance in the atmosphere, carbon dioxide (CO_2) is the most important anthropogenic greenhouse gas. Its global cycle can be divided into three main compartments: atmosphere, terrestrial biosphere, and ocean, see Figure 1.1. Natural processes, such as photosynthesis, plant and soil respiration,

CHAPTER 1. INTRODUCTION

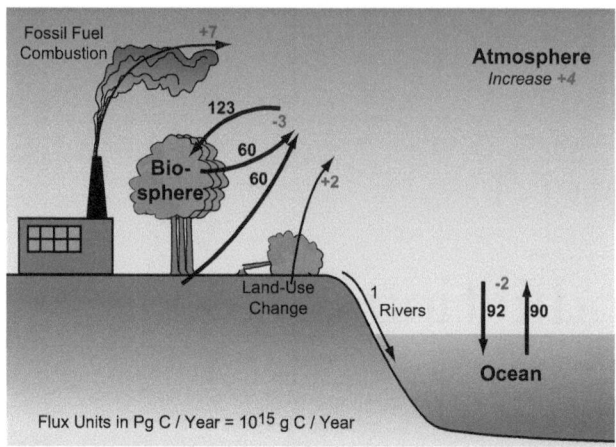

Figure 1.1: Overview of the global carbon fluxes (after IPCC, 2007; Levin, 2009). (Gross fluxes are marked in black and net fluxes in red.)

or sea surface exchange, lead to massive fluxes between the terrestrial biosphere and the atmosphere ($\sim$ 120 PgC) and between the ocean and the atmosphere ($\sim$ 90 PgC). The two major sources of anthropogenic CO_2 emissions are from fossil fuel combustion ($\sim$ +7 PgC) and land-use change ($\sim$ +2 PgC). Not all of the emitted CO_2 stays in the atmosphere. The ocean and the terrestrial biosphere currently act as net sinks of carbon and remove about half of the human-caused emissions. However, the remaining emissions are responsible for a more than 75% increase in atmospheric CO_2 since the pre-industrial times (IPCC, 2007).

Shown in Figure 1.2 is the development of atmospheric CO_2 sources and sinks over the last 50 years. Although CO_2 emissions from fossil fuel combustion (gray shading) show a steady increase, the increase of CO_2 in the atmosphere (light blue shading) through this period has been highly variable. This variability is mainly due to changes in the magnitude of the terrestrial biosphere sink (green shading). The high variability can be largely attributed to the interannual variability of climate and demonstrates the dynamic role of terrestrial ecosystems in the global carbon cycle. The magnitude of this sink is essentially the integral over the net carbon fluxes of all terrestrial ecosystems, with each ecosystem responding individually to the local climatic conditions. Detailed knowledge of the carbon fluxes between the terrestrial ecosystems and the atmosphere is thus fundamental for understanding the global carbon cycle and predicting the

CHAPTER 1. INTRODUCTION

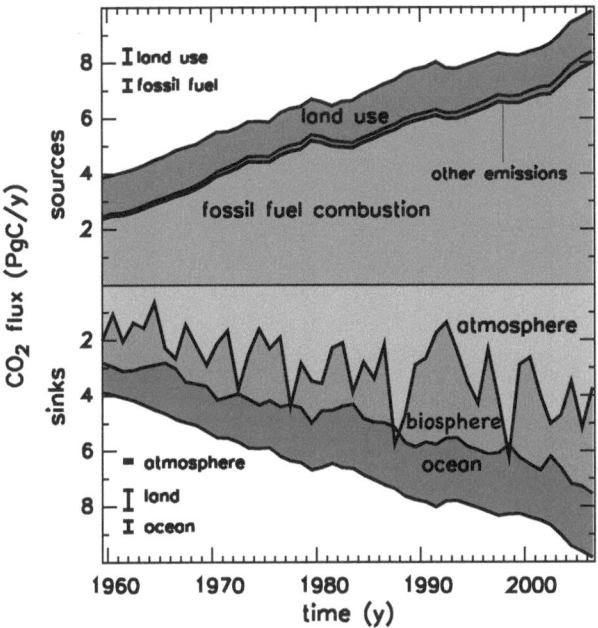

Figure 1.2: Sources and sinks of atmospheric CO_2 from 1959 to 2006. *Top:* The sources of CO_2 emissions as the sum of fossil fuel combustion, land-use change, and other emissions. *Bottom:* The sinks of CO_2 in the biosphere and in the ocean and the net overall increase in CO_2 concentration in the atmosphere. Figure taken from Canadell *et al.* (2007).

effects of climate change.

1.2 Data domain

1.2.1 Terrestrial carbon flux measurements

With the development of the eddy covariance technique, carbon, water, and energy fluxes between the terrestrial ecosystems and the atmosphere can be directly measured at the ecosystem level (Baldocchi, 2008). Eddies are the turbulent motions of upward and downward moving air which transport gases such as CO_2. These turbulent motions and the concentration of the gases within

CHAPTER 1. INTRODUCTION

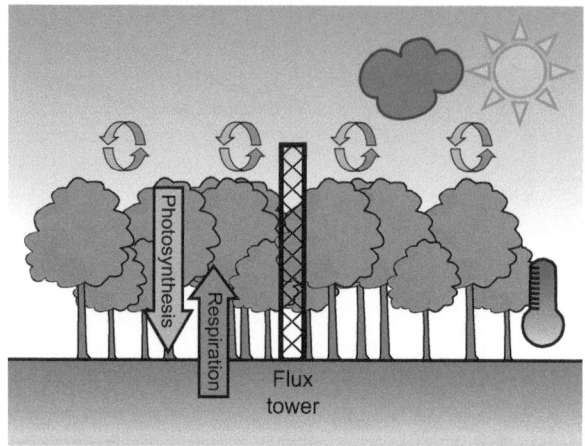

Figure 1.3: Sketch of flux tower measurements over a terrestrial ecosystem.

are sampled with a three-axis sonic anemometer and a gas analyzer. The vertical flux of CO_2 can be calculated from the covariance between the fluctuations of the vertical motion and the CO_2 concentrations. The measurements are taken with a frequency on the order of 10 Hz. To calculate mean flux densities (typically averaged over half-hourly or hourly time spans), several data treatment steps are required (Aubinet et al., 2003).

The eddy covariance instruments are installed at flux towers above the ecosystem canopy (see Figure 1.3). The source area of this flux, the footprint, has typical longitudinal length scales of 100 m to 2000 m (Schmid, 1994). The carbon flux measured is the net ecosystem productivity NEP, which is the photosynthetic uptake minus the release by respiration. In contrast to controlled lab experiments, the ecosystem response is driven by external weather conditions. To capture the climatic conditions, there are concurrent measurements of a wide range of variables, such as radiation, temperature and humidity.

Continuous routine applications of this meteorological technique became available since the early 1990s and eddy flux towers have now been established to monitor the CO_2 fluxes over a wide range of vegetation types and climate zones all over the world (Baldocchi, 2008). The carbon fluxes and the auxiliary meteorological data are usually recorded throughout the whole year with a high temporal resolution of half-hourly or hourly intervals.

CHAPTER 1. INTRODUCTION

Figure 1.4: Photo of the flux tower at Hainich forest.

1.2.2 Description of the Hainich forest tower site

One particular site with eddy flux measurements is the Hainich forest in Germany (Knohl et al., 2003). The Hainich tower, see photograph in Figure 1.4, is located at 51° 04' 46" N, 10°27' 08" E, and 440 m above sea level in one of the largest beech forests of Central Europe. Due to its history as a military base, the forest has been taken out of management for more than 70 years. In the centuries before, the Hainich forest was managed by the local village population in a sustainable coppice-with-standards method. Therefore, this forest developed basically undisturbed with trees covering a wide range of age classes up to 250 years. The forest is dominated by beech (*Fagus sylvatica*, 65%), ash (*Fraxinus excelsior*, 25%), and maple (*Acer pseudoplantanus* and *Acer plantanoides*, 7%). The forest floor is completely covered with understory vegetation (*Allium ursinum*, *Mercurialis perennis*, *Anemone nemorosa*). Details on the stand characteristics in the main footprint area of the tower are given in Table 1.1.

The typical phenology of the Hainich forest is: active understory vegetation

CHAPTER 1. INTRODUCTION

Parameter	Value
Stand density	334 trees/ha
Basal area	34.2 m^2/ha
Mean diameter at breast height	0.308 m
Mean tree height	23.1 m
Maximum tree height	37 m
Living tree biomass	21 270 g C/m^2
Carbon in organic layer	295 ± 26 g C/m^2
Carbon in mineral soil (0-50 cm depth)	12 220 g C/m^2
Dead wood	928 g C/m^2
Maximum LAI of canopy (2004)	5.5 m^2/m^2
Aboveground tree litter production (2004)	300 ± 52 g C/m^2

Table 1.1: General stand characteristics of the main footprint area of the Hainich forest tower site in Thuringia, Germany. Table taken from Kutsch *et al.* (2008).

from April to October, leafed trees from May to October, and a dormant season from November to March (with only sporadic snow cover). The climate is temperate suboceanic/subcontinental with long term annual means of 7.5°C - 8°C air temperature and 750 - 800 mm precipitation. The prevailing wind direction at the site is from the south-west.

1.2.3 Properties of the eddy flux measurements

Due to limitations of the eddy covariance technique and ecosystem measurements in general, the obtained datasets are subject to many sources of uncertainty:

Fragmentation: Instrumentation failure, stationarity test, spike filtering, footprint issues, horizontal advection flow, and other quality criteria and checks might lead to a rejection of the measurement (Goeckede *et al.*, 2004; Papale *et al.*, 2006). The main limitation of the eddy covariance technique is the requirement for turbulent atmospheric conditions. During the daytime, positive sensible heat fluxes create buoyancy that helps

CHAPTER 1. INTRODUCTION

to mix the atmosphere. At nighttime, however, radiative cooling leads to stable conditions that suppress turbulent mixing. Overall, the annual datasets are highly fragmented with 20% to 60% gaps in the data (Moffat et al., 2007). The majority of these gaps occur during nighttime.

Random error: The main sources of uncertainty are the noise in the measurement from the instruments as well as the turbulent transport. This random error has been assessed by Hollinger & Richardson (2005) by comparing carbon flux measurements from two towers with the same footprint; It scales with the magnitude of the flux and can be of the same order of magnitude as the measurements, especially during nighttime.

Multidimensionality: Since the ecosystem is driven by external weather conditions, the flux measurements are taken together with a wide range of auxiliary meteorological data, such as radiation, temperature, and humidity. Each of these measurements has its own intrinsic measurement noise or even gaps. The highly multidimensional datasets are complex with hidden non-trivial causalities due to cross-correlations, confounding effects, and time lags (memory) of the ecosystem response.

Inconsistency: Even under the same meteorological conditions, the measured net carbon flux might be different due to changes in the state of the ecosystem, such as the phenology, soil properties, time lag effects, changes in the footprint, or missing additional meteorological drivers.

Abundance: The high temporal resolution leads to frequent recurrences in the dataset with the response mainly driven by the daily cycle. The underlying longer term changes only cause small effects. Because of this abundance, measurement gaps even of several days can be reconstructed without causing much additional uncertainty in the (gap-filled) annual sums, unless the ecosystem is in a period of active change such as springtime for a deciduous forest (Richardson & Hollinger, 2007).

Overall, the carbon flux datasets are so large, complex, and multi-dimensional that the causalities cannot be derived by visual inspection.

1.3 Methodological concept

1.3.1 Data problem class

Since the ecosystem is in a natural state (with history) driven by external weather conditions, the observed ecosystem response is highly complex with many coexisting and coupled causes and effects. Even high-resolution multi-dimensional measurements restrict the observer to snap-shots of certain aspects of the ecosystem response. The obtained ecosystem datasets usually exhibit the following properties:

- Fragmentary,
- Noisy,
- Multi-dimensional,
- Inconsistent,
- Abundant.

Eddy covariance measurements are a typical instance of these kinds of observational datasets. To interpret ecosystem datasets and understand the underlying causalities, a mapping of the ecosystem's behavior into mathematical models becomes necessary.

1.3.2 Inductive modeling approach

In general, two basic modeling approaches can be distinguished: the hypothetic-deductive and the inductive (Hempel & Oppenheim, 1948; Young & Jarvis, 2002).

The hypothetic-deductive approach (Figure 1.5, top) begins with hypotheses about how the controls in the ecosystem work. The controlling processes are then implemented in an ecosystem model as parameterized equations (deduction). The observational datasets are used to constrain the parameters and to test the validity of the model. A good agreement of the model's predictions with the measurements is assumed to corroborate the hypotheses.

The methodology presented here is an inductive approach (Figure 1.5, bottom) where a priori assumptions are avoided as much as possible. It is based on artificial neural networks (ANNs), a purely empirical model with a very general function class. The characterization of the ecosystem response to its climatic

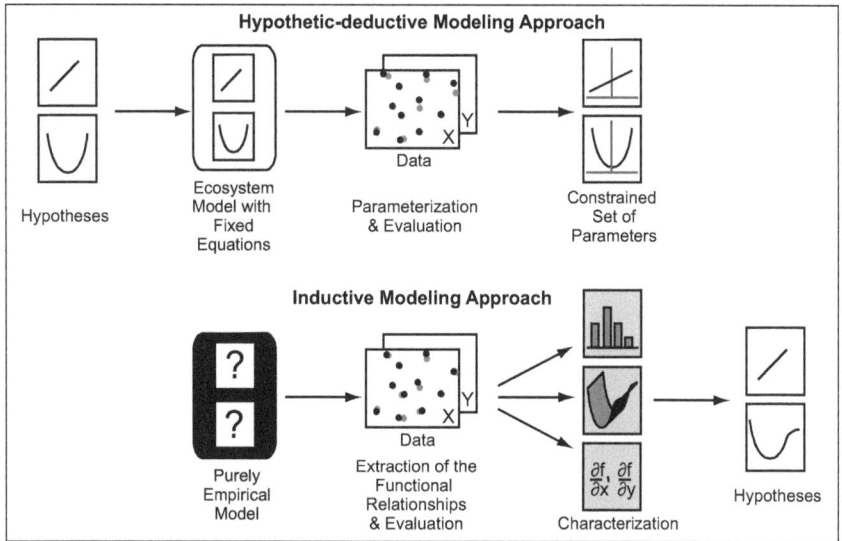

Figure 1.5: Conceptual flow of the two modeling approaches. (The shaded areas depict the special features of the inductive methodology presented here to characterize the ecosystem datasets.)

drivers is inferred solely and directly from the observations. Only at the last step of the inductive approach are the results put in the context of current hypotheses.

The inductive, data-derived modeling approach is usually diagnostic, i.e. the modeled response is an instantaneous response to the input forcing, see Figure 1.6. In contrast, hypothetic-deductive models can range from simple diagnostic to highly complex prognostic models. The latter are conceptional descriptions of the biological and physical processes also accounting for the ecosystem state, i.e. the memory of the natural system. In this thesis, synthetic data from two types of hypothetic-deductive models will also be considered: simple non-linear regressions (NLRs) and prognostic terrestrial biosphere models (TBMs).

1.3.3 Characterization with artificial neural networks

The strength of ANNs as the underlying inductive framework is their ability to recognize correlations between the drivers and the ecosystem response even

CHAPTER 1. INTRODUCTION

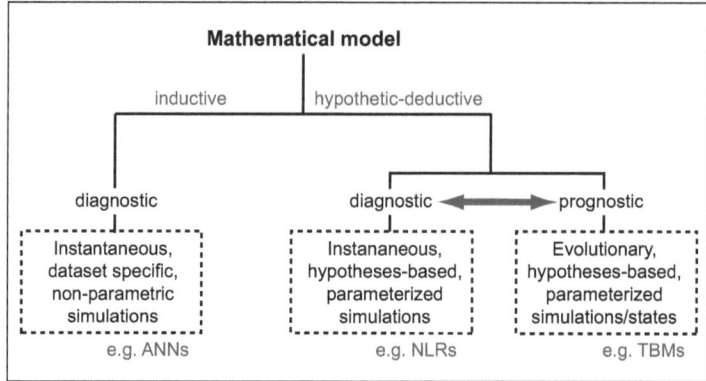

Figure 1.6: Mathematical model types for representing the ecosystem response.

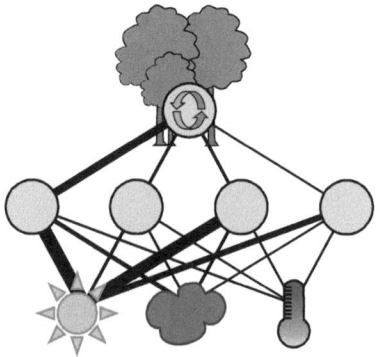

Figure 1.7: Schematic of an ANN model trained on eddy flux measurements with the three climatic drivers radiation, humidity, and temperature.

in the highly complex and noisy observational datasets described above. The ANN model trained on ecosystem data represents the instantaneous ecosystem response to the climatic controls as sketched for eddy flux measurements in Figure 1.7.

ANNs are usually used for spatial or temporal interpolation in a "black box mode", i.e. without further examination of the mapped correlations. The work presented here explores the capability of ANNs to be used in an inductive manner as a "glass box". By analyzing the purely empirical relationships mapped by the ANN models from the data, the ecosystem response to its climatic drivers

CHAPTER 1. INTRODUCTION

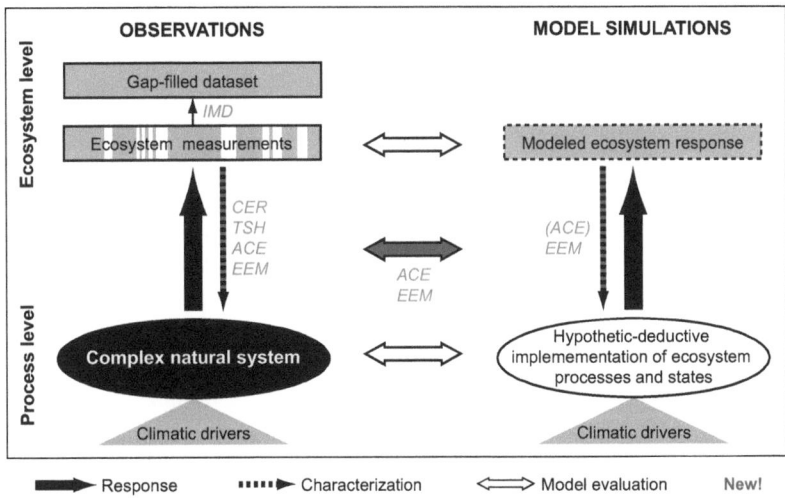

Figure 1.8: Observation and model simulation streams at the process and ecosystem levels. (Noted in gray italics are the corresponding areas of application of the methodology.)

can be characterized. The characterization ranges from benchmarking the total explainable variability and identifying the hierarchy of the controls to extracting the multivariate dependencies and sensitivities (see also the shaded areas in Figure 1.5, bottom).

1.3.4 Basic principle

The methodology follows an inductive modeling approach: very general models are generated and then systematically specialized on (subsets of) the ecosystem data in such a way that the specialized models can be interpreted ecophysiologically. The general model class is based on ANNs, specifically implemented to support a wide range of mathematical characterizations of the underlying correlations. The different phases of the methodology describe a framework to extract the underlying functional relationships with a minimum of prior information. This permits a characterization of the ecosystem response to its climatic drivers directly from data.

The context of the methodology is illustrated in Figure 1.8, sketching the observation and model simulation streams at the process and ecosystem levels: The ecosystem measurements can be seen as point observations of the ecosystem response to the climatic drivers depending on the history and state of the ecosystem. The modeled ecosystem response is the result of interactions of various implemented processes to the climatic drivers in dependence on the initial conditioning and modeled states.

The standard use of the observations and model simulations is forward (thick black arrows). In contrast, the presented methodology allows a new inverse characterization of the observed as well as the modeled ecosystem response to its climatic controls (thin dotted arrows). Model evaluation is usually performed at the process level or at the ecosystem level (white double arrows). The inverse characterization also provides a new model evaluation scheme (red double arrow): a link between the observations and their representation in the modeling world.

With the inverse characterization of the observational or synthetic data, the methodology opens a whole suite of novel ecophysiological applications:

1. Characterizing ecosystem responses (CER);
2. Testing specific hypothesis (TSH);
3. Assessing competing semi-empirical equations (ACE);
4. Evaluating ecosystem models (EEM); and
5. As a by-product: interpolating missing data (IMD).

Each of the five areas of application follows the phases of the methodology. The wide range of applications will be elaborated on the same ecosystem dataset throughout this work: the carbon flux measurements from the Hainich forest.

1.4 Outline of the thesis

The methodology presented in this thesis aims to provide a generally applicable systematic approach to extract the causalities hidden in ecosystem datasets with as little prior information as possible.

The manuscript is divided into three parts. Part I of the thesis describes the inductive methodology itself: the modeling framework (Chapter 2), the techniques employed to analyze the datasets (Chapter 3), and the seven phases of the methodology (Chapter 4).

CHAPTER 1. INTRODUCTION

Part II aims to demonstrate that the methodology is generally applicable to ecosystem datasets which exhibit the properties described above and, in particular, that the methodology is well-suited to investigate a wide range of different ecophysiological aspects in these datasets. The methodology will be elaborated on and its ecophysiological relevance examined for five different queries to cover each area of application: the daytime CO_2 response during the active season is used as the example for characterizing ecosystem responses (Chapter 5); the net effect of diffuse light is tested as a specific hypothesis (Chapter 6); seven competing equations are assessed in their performance as light response curves (Chapter 7); evaluating ecosystem models is examined for two terrestrial biosphere models (Chapter 8); and missing data in the dataset is interpolated, the so called gap-filling (Chapter 9).

In the final part, Part III, the results of the application examples are summarized with regard to the reliability of the modeling results (Chapter 10), followed by general conclusions and outlook (Chapter 11). Details on the technical implementation are presented in the appendix (Appendix A).

CHAPTER 1. INTRODUCTION

Part I

Methodology

Chapter 2
Modeling framework

An inductive approach requires a purely empirical modeling framework. Purely empirical models are broadly used, e.g. for the spatial or temporal interpolation of the carbon fluxes (Gove & Hollinger, 2006; Papale & Valentini, 2003; Stauch & Jarvis, 2006), but mostly in a black box mode. Only a few of them are employed in an inductive manner also aiming to provide a physiological interpretation, such as data-based mechanistic modeling by Young & Jarvis (2002).

The purely empirical model used here is based on statistical multivariate modeling with artificial neural networks (Bishop, 1995; Rojas, 1996). An ANN consists of an interconnected group of artificial neurons, also called nodes, which can be trained to learn the correlations present in the dataset. The methodology exploits their outstanding data-mining ability, i.e. their ability to recognize the underlying patterns even in large sets of (noisy) observational datasets. ANNs have been shown to outperform classical semi-empirical methods (e.g. Abramowitz, 2005; Moffat *et al.*, 2007) and can thus be used as a benchmark for process-based model descriptions (Abramowitz, 2005).

This chapter describes the modeling framework: how the very general models are generated and how the correlations of the target variable (response) with the controlling input variables (drivers) are learned. The type of ANNs used here are feedforward ANNs (Section 2.1). The learning is based on the backpropagation (BP) algorithm (Section 2.2). In this algorithm the partial derivative of the error is backpropagated through the network. The same principle can be used to also calculate the partial derivative of the network function (Section 2.3). The nodes have a logistic activation function and optimal learning requires scaling of the training data (Section 2.4). The actual learning of the network, the training process, follows a complex procedure including pruning of the nodes and smoothing towards the minimum error (Section 2.5).

2.1 Feedforward artificial neural networks

A feedforward ANN consists of nodes interconnected by weights only in a directed forward way. Information thus moves from the input node layer forward through the hidden node layer(s) to the output node layer. This type of ANN provides the features required by this inductive approach: a very general function class with a closed-form expression of the network, trained by a supervised learning algorithm that is suited to non-linear regression tasks. Cybenko (1989) proved that a single hidden layer, feedforward ANN is capable of approximating any continuous, multivariate function to any desired degree of precision. This means that a complex enough feedforward ANN has the built-in flexibility to map the individual conditions without needing prior assumptions about the shape of the response.

2.2 Backpropagation algorithm

The backpropagation algorithm is a widely employed method for supervised learning of feedforward ANNs (Rojas, 1996; Rumelhart *et al.*, 1986). A supervised learning process requires essentially the same three choices as other model-data synthesis processes (after Raupach *et al.*, 2005):

1. A sufficient model with adjustable properties,
2. A measure of the distance between data and model, also called the cost or error function, and
3. A search strategy for finding an optimum of the cost function.

2.2.1 Network topology

An ANN model can be described by its network topology and represented in a diagram as shown in Figure 2.1. The number of input and output nodes of the feedforward ANN are set up according to the dimensions of the data used for training. The number of hidden layers and the number of nodes within each hidden layer can vary from one up. The weights determine the behavior of the

CHAPTER 2. MODELING FRAMEWORK

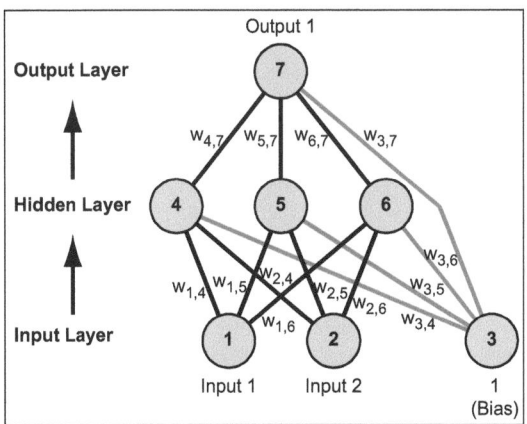

Figure 2.1: Example of a feedforward ANN with two input nodes plus the bias node, three nodes in the single hidden layer, and one output node.

network and are the model properties to be adjusted during training. The extra weight of each node connected to a constant, here 1, is the bias of the node.

For the model to be sufficient, it has to have enough degree of freedom to be able to learn the training tasks. The degree of freedom of an ANN depends on the number of weights and can be changed by adding or pruning nodes or hidden layers.

Adding or pruning a hidden layer affects all weights to the layer above and below and results in an erratic change in the behavior of the ANN model, whereas adding or pruning of single nodes shows more gradual effects. Furthermore, going from a one layer to a two-layer network greatly increases the degree of freedom of the network, the training process takes a lot more computing time, and the pruning of the nodes is less efficient. Since Cybenko (1989) showed that one hidden layer is sufficient, a single hidden layer network with a large enough number of nodes[1] was used as the initial network topology throughout this work. For the presented application examples, the number of initial hidden nodes (before pruning) was in the range of six to ten.

---

[1] If the function approximation did not improve further by adding more nodes, the number of hidden nodes was regarded as *large enough*.

2.2.2 The error function

In the case of the backpropagation algorithm, the error function E to be optimized is the Euclidean distance. For a single output node, the equation is:

$$E(\mathbf{W}) = \frac{1}{2} \sum_{i=1}^{N} (m_i - t_i)^2, \tag{2.1}$$

where t_i are the individual (observed) target values, m_i are the values modeled by the ANN, and N is the total number of data tuples in the training dataset. $\mathbf{W}$ is the weight matrix and its elements $w_{k,l}$ denote the connection weight between node k and node l.

The gradient descent optimization is used as the search strategy to minimize the error function, i.e. as the training algorithm for the adjustable network weights. The derivative of the error function is propagated back through the network following the steepest descent with respect to the weights. Each weight $w_{k,l}$ is updated using an increment:

$$\Delta w_{k,l} = -\gamma \frac{\partial E}{\partial w_{k,l}}, \tag{2.2}$$

where γ is the learning rate. The weight update $\Delta w_{k,l}$ is also called the delta rule. The learning rate γ is a positive proportionality parameter which defines the step length of each iteration in the negative gradient direction. The combination of weights which minimizes the error function is considered to be a solution of the learning problem.

2.2.3 Online and epoch mode

The backpropagation training algorithm can be executed in two modes:

3a. Online: the data tuples arrive in sequence and are processed one by one, also called sequential or stochastic mode;

3b. Epoch: the data tuples are processed all at once, also called offline, non-sequential, or batch mode.

In the online case, the weights are updated after each data tuple presentation and so the weight updates do not follow the true gradient exactly. However, if the data tuples of the training dataset are randomly presented, the network learning oscillates around the true gradient direction. Online training is more efficient when the number of data tuples is large. Therefore, the training algorithm is usually executed in online mode on the eddy flux datasets.

An epoch is one round of all data tuples presented to the network. The sum of the gradients over all data tuples of the training dataset is used to update the weights, following the true gradient. This mode is used to finalize the training process smoothly.

2.2.4 Momentum term

To avoid convergence problems of this algorithm in "steep gullies" or on "flat plateaus" of the error function, a momentum term is added to provide the search process with a kind of inertia:

$$\Delta w_{k,l} = -\gamma \frac{\partial E}{\partial w_{k,l}} + \eta \Delta_{-1} w_{k,l}, \qquad (2.3)$$

where η is the empirical momentum term and $\Delta_{-1} w_{k,l}$ the weight update value of the previous step.

2.2.5 Regularization term

If the trained ANN models tend to overfit fine details of the training data, the implementation of a weight regularization algorithm can be beneficial (MacKay, 2003). The error function E in Equation 2.1 is extended by a term that penalizes large weights:

$$E'(\mathbf{W}) = \frac{1}{2} \sum_{i=1}^{N} (m_i - t_i)^2 + \alpha \cdot \frac{1}{2} \sum_{k=1}^{L} \sum_{l=1}^{L} w_{k,l}^2, \qquad (2.4)$$

where L is the total number of nodes in the network and α is a positive regularization constant, also called the weight decay rate.

The new delta rule for the weight update is:

$$\Delta w_{k,l} = -\gamma (\frac{\partial E}{\partial w_{k,l}} + \alpha w_{k,l}). \qquad (2.5)$$

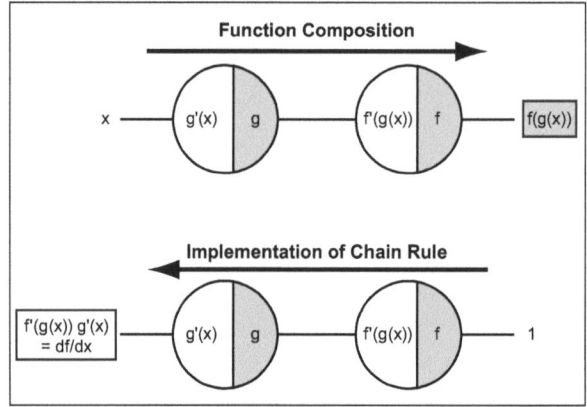

Figure 2.2: Backpropagation diagram for the feedforward step (top) and backpropagation step (bottom) after Rojas (1996).

The weight update for combined regularization (Equation 2.5) and momentum (Equation 2.2) is:

$$\Delta w_{k,l} = -\gamma(\frac{\partial E}{\partial w_{k,l}} + \alpha w_{k,l}) + \eta \Delta_{-1} w_{k,l}. \tag{2.6}$$

2.3 Partial derivative propagation

In the backpropagation algorithm, the partial derivatives of the error function towards the weights are propagated backwards through the network. This algorithm can be extended to also compute the partial derivatives of the network function. This extension can best be explained using a backpropagation diagram, a graphical approach after Rojas (1996). The BP diagram in Figure 2.2 shows the basic principle:

Each node consists of a left and right side. At the right side (gray), the node function is calculated, whereas the left side (white) stores the derivative of this function. In the feedforward step (top), the input value x is propagated through to the right side of the network diagram to compose the network function. In the backpropagation step (bottom), the derivatives are propagated through to the left side and the chain rule of the composite function is implemented.

In classical backpropagation, the inputs are treated as constants and the

CHAPTER 2. MODELING FRAMEWORK

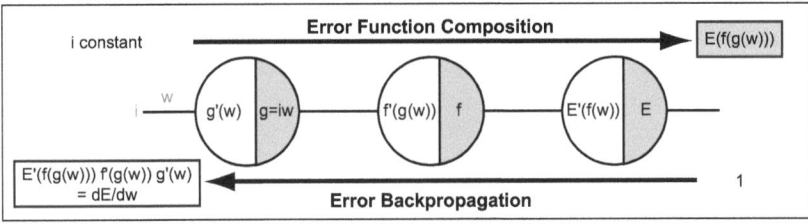

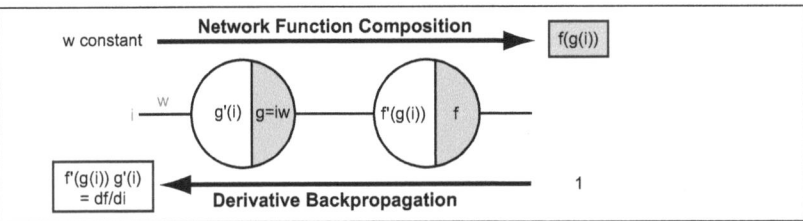

Figure 2.3: Error function composition and backpropagation (top), and network output function composition and backpropagation (bottom).

partial derivative of the error function with respects to each weight is propagated back through the network, see Figure 2.3, top. The derivative propagation of the network output function follows the same principle, only reversed with respect to inputs and weights. This time the weights are treated as constant and the partial derivative of the network output function with respect to each input is propagated back through the network, see Figure 2.3, bottom. The numerical partial derivatives are calculated for each data tuple.

The derivative propagation of the network output function was implemented by extending the nodes to also store the derivative of the node function. The backpropagation of the delta rule and the node derivatives are carried out simultaneously. This extension of the classical backpropagation algorithm can be generally incorporated into feedforward networks.

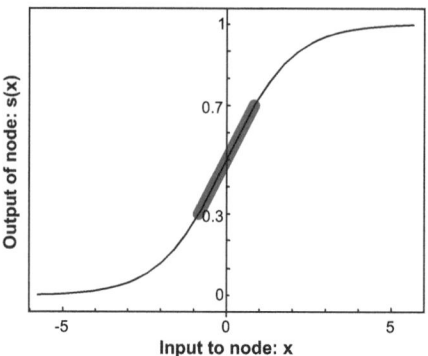

Figure 2.4: Graph of the logistic sigmoid activation function. (Highlighted in red the nearly linear range of the slope.)

2.4 Node properties

2.4.1 Activation function

The nodes' output function is called the activation function. For the computation of the gradient, this activation function has to be continuous and differentiable. A popular non-linear activation function for backpropagation networks is the logistic sigmoid $s(x)$:

$$s(x) = \frac{1}{1+e^{-x}}. \tag{2.7}$$

For high negative values, $s(x)$ is zero (deactivated), see Figure 2.4. With increasing positive signal strength from the incoming nodes, $s(x)$ of each node can be activated during training. In the learning process, the individual nodes are activated or deactivated through the adjustment of weights connecting the nodes. The derivative of the logistic sigmoid is very simple and efficient to compute, because it can be expressed as a polynomial of itself:

$$\frac{ds(x)}{dx} = s(x)\left(1 - s(x)\right). \tag{2.8}$$

CHAPTER 2. MODELING FRAMEWORK

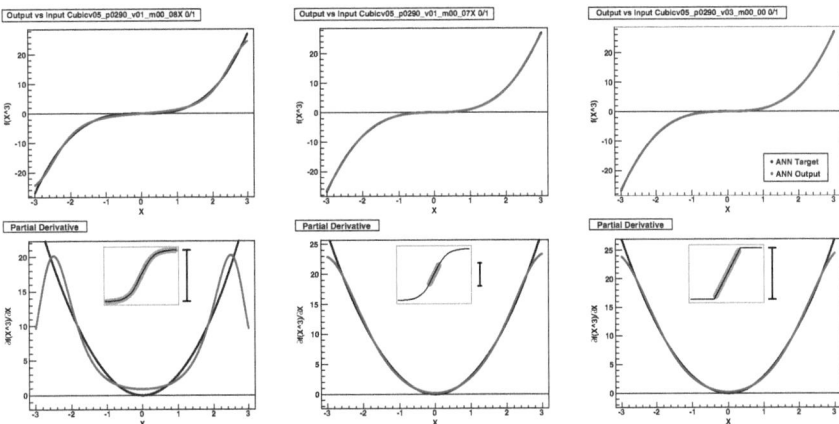

Figure 2.5: Modeled cubic function $f(x) = x^3$ (top) and derivative (bottom) for three different setups of activation functions of the output node and/or scaling range of the training pattern, as depicted in the small sketch. (The plot title contains the plot description, the pattern-specific ID, the version number of the training setup, the processing routine, and the training permutation, see Appendix A for further explanations.)

2.4.2 Scaling of the training pattern

The set of all tuples of input and target data used to train or run an ANN is called a pattern. To ensure an optimum training procedure of an ANN, the input and target values of a pattern need to be scaled to the operative range of the nodes' activation functions.

The inputs were scaled to [-1,1]. This range is within the nearly linear range of the logistic sigmoid and well outside the saturated areas of $s(x)$, see Figure 2.4. The different input values are thus weighted evenly.

For the scaling of the target values, the activation function of the output nodes has to be considered. For the logistic sigmoid, the targets are usually scaled to [0,1], the full output range (Lek & Guégan, 2000). This means that the saturated regions of the function would have to be activated to output values close to the lower or upper limit and that 0 or 1 are only reached at $-\infty$ or $+\infty$, respectively (see also Looney, 1997). The effect on the training is shown for the example of the cubic function in 2.5, left. The scaling of the targets with the full range leads to a large distortion of the derivative at the edges, where

25

$s(x)$ is saturated.

The best results were obtained, when the scaling was limited to the nearly linear range of the output of the sigmoid activation of [0.3,0.7]. Now the target values are weighted evenly and the edge effect is only small, see Figure 2.5, middle. A similar improvement can be reached using a piecewise linear function as the activation function of the output node, see Figure 2.5, right.

The scaling effect only became apparent when looking at the derivatives. The common [0,1]-scaling seems sufficient if the ANN models are used only for function approximation. The error of the trained ANN, thus the mismatch between the ANN output and target values, was small for all three setups in the presented example. However, this modeling framework is also used to look at the network function and its partial derivatives. Therefore, the fine-tuning step of scaling the target pattern to a restricted range of [0.3,0.7] is critical, especially since the mapped relationships are also interpreted for their ecophysiological properties.

2.4.3 Goodness factor

The goodness factor G_j describes the total signal propagated forward by the jth node (Murase *et al.*, 1991):

$$G_j = \sum_{i=1}^{N} \sum_{o=1}^{O_j} \left(w_{j,o}\ o_{j,i}\right)^2, \qquad (2.9)$$

where N is the total number of data tuples in the training pattern, $o_{j,i}$ is the value of the output function for each data tuple i, and O_j is the number of outgoing (forward) weights $w_{j,o}$ of the jth node. This squared sum over all signals that the other nodes receive from node j gives a measures of the activation of node j and can be used as diagnostic tools for network pruning.

2.5 Training process

After setting up a feedforward network and scaling of the training pattern, the actual training process is started. One round of presenting and learning of all training data tuples is called an iteration. After each iteration, several error measures, such as correlation coefficient R^2, the root mean square error $RMSE$,

CHAPTER 2. MODELING FRAMEWORK

and the bias error *Bias*, are calculated. (More details on the error measures can be found in Section 3.1.)

The ANN training process goes through several phases:

1. First, a Monte Carlo technique is used, where repeatedly networks with a randomly assigned initial weight configurations are generated. The configuration with the lowest *RMSE* and highest R^2 is used as the starting weight configuration.

2. Then the ANN is trained with the backpropagation algorithm in online mode with a high learning rate.

3. As soon as the *RMSE* training error levels off and improves less than a certain threshold (usually 1%), the training process is stopped (early stopping).

4. Next the pruning process is started: The node of a hidden layer with the lowest goodness factor (Section 2.4.3) is pruned, if its activation is a certain factor lower than of the node with the highest goodness factor. The training is continued with a low learning rate, until the *RMSE* levels off again. If the node was of little significance, the pruned network is likely to still be close to the last minimum and the training error should reach the same low error level as before. The pruning step is repeated, as long as there are nodes with a low enough activation to be pruned and as long as the *RMSE* recovers. If the training of the pruned network levels off to a higher error, thus a decreased performance, the last pruning step is undone.

5. The ANN training is finalized with epoch mode iterations. Since the epoch mode iterations follow the true gradient, the training progression is very smooth and settles towards the closest local optimum. This smoothing procedure turned out to be necessary to minimize the bias error.

6. The feedforward ANN is now fully parameterized and can be used to perform the data analysis. The trained ANN with fixed weights will be referred to as *the* ANN model.

Figure 2.6 shows a typical example of a training progression: First, a good initial weight configuration was located with a Monte Carlo technique. Then

CHAPTER 2. MODELING FRAMEWORK

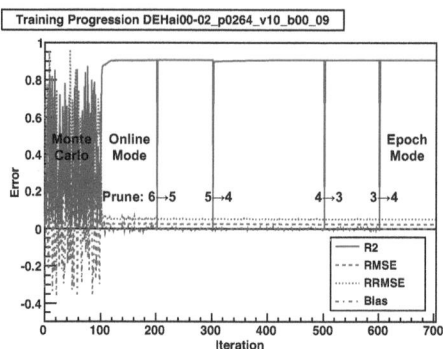

Figure 2.6: Typical example for an ANN training progression.

the ANN was trained in online mode. Once the *RMSE* leveled off, the pruning of the six nodes in the hidden layer started. The last pruning from 4 to 3 nodes resulted in a slightly higher error (not visible on the graph), and therefore the network with four nodes was reloaded. The ANN training was finalized in epoch mode. The example is actually the training progression of an ANN model used in Section 5.5.2 with the final network structure provided in Figure 5.13.

In the training process, the starting weights are randomly assigned and the order of the data tuples learned are randomly shuffled within each iteration. Therefore each ANN training, even on the same dataset, has a different training progression and thus yields a differently trained ANN model with a different final node and weight configuration. However, the ANN models trained on the same dataset should optimally yield a similar performance and function approximation. To get a measure of the robustness of the trained ANN models, the training process was repeated ten times throughout this work.

The presented modeling framework provides a general basis to generate and train a purely empirical model, the prerequisite of an inductive methodology. The reliability of the framework, such as the generalization beyond the training dataset or the robustness of the training permutations, is discussed later in Chapter 10 with reference to the modeling results presented in Part II of the thesis. How this inductive modeling framework can be used to analyze ecosystem datasets, is the subject of the next chapter.

Chapter 3
Data analysis toolbox

The modeling framework introduced in the last chapter can now be employed to train feedforward ANNs on ecosystem datasets. Since the dataset is presented as snapshots, one tuple at a time, the purely empirical models pick up correlations at the presented time scale. After successful training, the ANN model maps the underlying correlations of the responding output variable to the controlling input variables (drivers). The properties of the modeled response are then used to identify the mathematical characteristics of the dataset it was trained on. In the following, several tools to analyze the data are described:

The mapping performance gives a measure of how much of the response can be explained with the provided input drivers (Section 3.1). If used on driver subsets, the mapping performance can also be used to identify the relevance of the drivers (Section 3.2). The analytical network function describes the functional relationship of the response to the drivers (Section 3.3), while its (numerical) partial derivatives characterize the response to changes in the drivers (Section 3.4).

3.1 Mapping performance measures

The quality, or performance, of the ANN model can be used to estimate how much of the response can be mapped (explained) with the input drivers provided. There are several performance measures of interest:

During the ANN training, the Euclidean distance E is optimized (see also Section 2.2). E is equal to twice the sum of squared errors (SSE_{err}) and twice the squared root mean square error ($RMSE^2$):

$$E = \frac{1}{2}\sum_{i=1}^{N}(m_i - t_i)^2 = 2 \cdot SSE_{err} = 2 \cdot RMSE^2, \qquad (3.1)$$

where t_i are the individual (observed) target values, m_i are the values modeled by the ANN, and N is the total number of data tuples.

The coefficient of determination (nlR^2) is directly related to SSE_{err} normalized by the total variance SSE_{tot} of the dataset:

$$nlR^2 = 1 - \frac{SSE_{err}}{SSE_{tot}} = 1 - \frac{\sum(m_i - t_i)^2}{\sum(t_i - \bar{t})^2}, \qquad (3.2)$$

where $\bar{t}$ is the mean of the (observed) target values. Since nlR^2 is directly related to SSE_{err}, and $RMSE$ respectively, and since SSE_{tot} is constant for the same dataset, nlR^2 provides a universal measure of the correlation as well as the model error.

Assuming a linear relationship between the modeled and (observed) target data, this coefficient of determination can be expressed as the squared correlation coefficient (R^2):

$$R^2 = \frac{\{\sum(m_i - \bar{m})(t_i - \bar{t})\}^2}{\sum(m_i - \bar{m})^2 \sum(t_i - \bar{t})^2}, \qquad (3.3)$$

where $\bar{m}$ is the mean of the (observed) target values. nlR^2 is a measure of the unexplained variance and can range from [-∞,1], whereas R^2 describes purely the correlation between two variables and has a fixed range of [0,1]. Since the SSE_{err} is optimized during the ANN training, these two measures are the same for the data tuples of the training pattern and therefore only R^2 is stated throughout this manuscript. However, if the analysis is performed on a different pattern than the training pattern or a subset, nlR^2 and R^2 might differ.

If the data follows a Laplace rather than a Gaussian distribution, not the $RMSE$ but the standard deviation $SDev$ should be used as a measure of the model residuals (Richardson et al., 2006; Lasslop et al., 2008):

$$SDev = \sqrt{2} \cdot MAE = \sqrt{2} \cdot \frac{1}{N}\sum|m_i - t_i|, \qquad (3.4)$$

where *MAE* it the mean absolute error.

For ANN models that will be used for predictions, the mean bias error (*Bias*) should also be considered (Moffat *et al.*, 2007):

$$Bias = \frac{1}{N} \sum (m_i - t_i). \qquad (3.5)$$

As a convention for this manuscript: If the ANN model has high correlation measures (nlR^2 and R^2 close to 1) and low errors (e.g. *RMSE*, *SDev*, or *Bias* close to zero), its mapping performance P will be referred to as *high*. Accordingly, a performance *improvement* implies increased correlation and decreased errors.

3.2 Driver relevance

The mapping performance P_i of an ANN model with a single input driver d_i measures how much of the response can be mapped by d_i. This can be used to determine the relevance of the drivers:

If a driver variable d_1 has more direct correlation with the responding output variable than another driver variable d_2, the mapping performance P_1 of the ANN model with single d_1 will be higher than the mapping performance P_2 of the ANN model with single d_2:

$$P_1 > P_2. \qquad (3.6)$$

The ANN mapping performance with single inputs can thus be used to quantify their importance as primary input drivers.

3.2.1 Added performance

In the same manner, the improvement in ANN performance with a new driver added to an existing network can be used as a measure of importance as an additional driver. The more *new* information the additional driver d_A adds, the greater is the improvement in the network performance, and the more relevant is this climatic variable d_A for the response (van de Laar *et al.*, 1999). The performance improvement ΔP_A can be calculated as:

$$\Delta P_A = P_{+A} - P, \qquad (3.7)$$

where P is the performance without d_A, and P_{+A} is the performance with d_A added.

When using the performance improvement as a measure of relevance, attention has to be paid to correlations between the input drivers. If a new driver adds little information to the system, it might mean that it is not relevant but it could also mean that the information is already present in the existing inputs. This fact can be used to detect correlations by first training the networks separately on two drivers of interest, and then together. Assuming that the two drivers showed high relevance when trained separately but only little added performance when both were used for the training, it means that the two are closely (cor-)related.

3.2.2 Greedy search algorithm

The added driver relevance can be employed in a search algorithm to determine the ranking of the drivers. A greedy search algorithm tries to reach the global optimum by keeping the local optimum fixed at each stage:

In the first step, the primary drivers of the response are identified by presenting ANNs with a single input variable at a time. The input variable yielding the highest ANN performance is the dominant driver of the response. In the second step, the ANNs are trained with the most relevant primary driver and one additional input variable at a time to determine the secondary drivers. The improvement in network performance is used to quantify the relevance of each added input variable. In a third step, the most relevant primary and secondary drivers are fixed to find the most relevant tertiary driver, and so on.

3.3 Network function

The ANN model maps the response of the dependent output variable(s) to the input drivers as present in the data. In a feedforward network, the input drivers d_1 to d_n are mapped unidirectionally from layer to layer onto the predicted output. This yields a unique, continuous analytical network function f describing the response:

$$f(d_1, ..., d_n), \text{ where } f : D \to \mathbb{R} \text{ and } D \subset \mathbb{R}^n. \qquad (3.8)$$

If the input drivers are mapped onto multiple outputs m, then f is a vector of $\mathbb{R}^m$, where each component represents a closed-form expression of the input dimension.

3.4 Numerical partial derivatives

The numerical partial derivative PaD of f with respect to each input driver d_i characterizes the change in the system response for each individual dataset tuple t_j:

$$PaD_{i,j} = \left(\frac{\partial f}{\partial d_i}\right)\bigg|_{t_j}. \tag{3.9}$$

The idea to also analyze the numerical partial derivatives was originally motivated by a comparison of techniques to determine the driver relevance by Gevrey *et al.* (2003). In this comparison, a method based on the sum of squared partial derivatives by Dimopoulos *et al.* (1999) was found to be the most useful.[1] In the work presented here, the sum of (un-squared) partial derivatives is used to look at the sensitivities of the response with respect to the drivers, as described below.

3.4.1 Dynamic range normalization

The numerical partial derivative $PaD_{i,j}$ describes the change of the response per measured physical unit. To be able to compare the partial derivates among drivers with different physical units or with different ranges, the numerical partial derivatives of each input driver d_i are linearly transformed from the dynamic range to unit range:

$$[d_{i,min}, d_{i,max}] \longmapsto [0, 1], \tag{3.10}$$

where $d_{i,min}$ is the driver minimum and $d_{i,max}$ the driver maximum the ecosystem is exposed to. As estimates of $d_{i,min}$ and $d_{i,max}$, the overall minimum and maximum of d_i in the annual dataset(s) is taken.

The normalized numerical partial derivative is calculated as:

$$nor.PaD_{i,j} = (d_{i,max} - d_{i,min}) \cdot PaD_{i,j}, \tag{3.11}$$

---

[1] Both papers compared the various ANN approaches also to multiple linear regression (classically used for ecological data) and found that the ANNs yield better results due to their capacity to take into account non-linear relationships.

CHAPTER 3. DATA ANALYSIS TOOLBOX

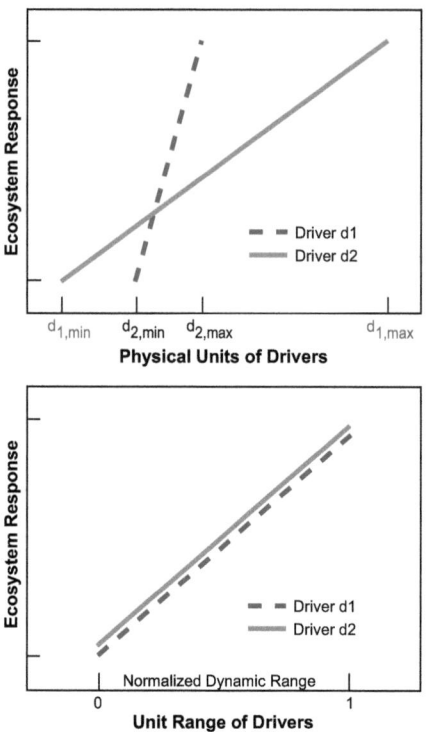

Figure 3.1: Sketch to illustrate the effect of dynamic range normalization.

which has the same scale for each of the drivers, namely units of system response per unit-normalized dynamic range.

A simple sketch to illustrate the normalization is given in Figure 3.1. The derivative (slope) of the response per physical unit to driver 1 is about five times higher than to driver 2 (top). Typical input drivers with similar proportions in the dynamic range are air and soil temperature. After normalization of the dynamic range to 1 (bottom), the change (slope) of the response with the respect to the two drivers is the same for this simple example of two linear dependencies.

3.4.2 Mean derivatives

To get an estimate of the mean absolute change of the response, the absolute numerical partial derivatives for an input variable d_i are averaged over all N tuples t_j:

$$abs.PaD_i = \frac{1}{N} \sum_{j=1}^{N} |nor.PaD_{i,j}|. \tag{3.12}$$

The positive and negative fractions of this sum provide information on negative and positive changes in the response:

$$neg.PaD_i = \frac{1}{N} \sum_{PaD_{i,j}<0} (nor.PaD_{i,j}) \text{ and} \tag{3.13}$$

$$pos.PaD_i = \frac{1}{N} \sum_{PaD_{i,j}>0} (nor.PaD_{i,j}). \tag{3.14}$$

This chapter described several tools to analyze the mathematical properties of the modeled response: the mapping performance, the driver relevance, the network function, and the partial derivatives. The different phases of the methodology presented in the following use the data analysis tools to characterize the ecophysiological relationships present in the dataset.

CHAPTER 3. DATA ANALYSIS TOOLBOX

Chapter 4
Phases of the methodology

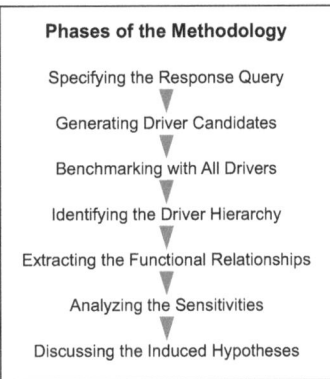

Figure 4.1: The seven phases of the methodology.

The methodology is a systematic approach based on seven phases. These phases guide to characterize the ecosystem response directly from complex and even fragmented datasets with as few prior assumptions as possible. The goal is to specialize the training of the purely empirical models (Chapter 2) on (subsets of) the dataset and to use the data analysis tools (Chapter 3) on these specialized models in such a way, that the modeled responses can be interpreted ecophysiologically.

The phases start with specifying the response query (Section 4.1) and generating driver candidates (Section 4.2). The data analysis tools are then used for benchmarking with all drivers (Section 4.3), identifying the driver hierarchy (Section 4.4), extracting the functional relationships (Section 4.5), and analyzing the sensitivities (Section 4.6) in the ecosystem dataset. As the last phase, the induced hypotheses are discussed in their ecophysiological context (Section 4.7).

4.1 Specifying the response query

Ecosystem datasets usually contain responding and driving variables. The response query can only comprise aspects of the ecosystem response that have their cause and effect hidden in the dataset and that can thus be extracted by the purely empirical modeling framework.

In the case of carbon flux datasets, the observed net carbon flux is the ecosystem response to its climatic drivers which are sampled by the auxiliary meteorological data. Additionally, there might also be some slower varying, descriptive state variables, such as the phenology, which are not direct drivers but nonetheless have an effect on the magnitude of the response. The analysis can be performed not only on the time scale of the measurements but also on aggregated (e.g. monthly) data.

To ensure that the dataset is representative for the characterization of the queried ecosystem response, the following pre-considerations should be taken into account:

Quality: Since an empirical model will map the functional relationships as present in the datasets, it is important to use accurate measurements that have integrity. The integrity of the dataset does not depend on the quantitative amount of data, but on the enclosed information contained. Therefore, the training datasets should only contain high quality data.

Timescale: The empirical model only maps the instantaneous response to the stimulus at the presented timescale. The time frame of the query thus needs to match the time scale of the dataset.

Data coverage: A purely empirical model can only properly map functional relationships within the scope of the training dataset. Interpolation between underrepresented regions or extrapolation might lead to physiologically implausible mapping. Therefore, the measurements used for training should have good data coverage over the full range of interest.

Reducing complexity: The more explicit the information in the dataset, the more concise will be the mapped relationships. To reduce the complexity of the modeled ecosystem response, only data relevant to the response query should be considered. For example, if the interest is in the photosynthetic response of the ecosystem, the dataset should be restricted to the daytime data of the active period.

A dataset satisfying the four items above can then be used to train the purely empirical models and extract the overall response in the dataset. To examine different aspects of the queried response, the dataset can be further grouped (training and analysis) or binned (analysis only) into subsets:

1. Data grouping: To investigate differences in the response, the representative dataset can be grouped into subsets (e.g. each month). The ANN models are trained separately for each subset, and the differences between them give insight into the variability of the response. The setup of the grouping can be varied to test for diurnal, seasonal, or interannual variability.

2. Data binning: To analyze certain aspects of the overall response, the representative dataset can be binned to certain variable ranges (e.g. carbon flux magnitudes). The ANN models are trained on the complete dataset, but the data analysis is performed on the individual bins.

The interpretation of the modeled response of these data subsets may lead to even further insight into the ecophysiological properties.

4.2 Generating driver candidates

The starting set of driver candidates should contain at least all driver variables that are assumed to have an effect on the queried ecosystem response. This is not restricted to the observable climatic drivers and should be extended to theoretical[1] variables that might add further relevant information. For example, the seasonal changes of the ecosystem can be included with latent[2] variables, such as the weekly mean temperature, or a fuzzy set[3] of variables for the course of the season (Papale & Valentini, 2003). Other latent variables, such as the fraction of diffuse light, might expose a different aspect of the response. Time

---
[1]not directly observable
[2]inferred from observations
[3]set of variables with gradually varying membership (Zadeh, 1965)

lag effects can be included by providing information about preceding events, such as previous productivity rates.

A comprehensive driver set provides a basis for multiple working hypotheses. The starting set of input drivers may include all available meteorological variables plus many theoretical drivers and the following phases can be used to determine whether these drivers have an actual relevance for the response or not. However, it is crucial not to miss relevant driver information, as this might distort the genuineness of the results.

4.3 Benchmarking with all drivers

The ANN model trained with *all* generated driver candidates provides a performance benchmark of the maximum mapping between the responding variable and the drivers. If the determination coefficient nlR^2 is used as the performance measure (Section 3.1), then the benchmark describes the total explainable variability in the dataset.

The unexplained variability can be due to not-included but relevant climatic controls, due to missing information about the state of the system, such as the phenology, and due to the noise in the measurement. Assuming an optimally trained ANN model with all relevant climatic drivers and sufficient information about the state of the system, the correlations of the drivers to the response at the chosen time scale will be fully mapped. The *SDev* of the model residuals can then be used to give an estimate of the uncertainty (random error) in the flux measurements (Richardson *et al.*, 2008). Another important application of the benchmark models is the interpolation of missing data, the so called gap-filling.

For little data and lots of input drivers, benchmarking with all drivers might lead to overfitting. To avoid this effect, the benchmarking phase can be repeated after the next phase of identifying the driver hierarchy using only the most relevant drivers.

4.4 Identifying the driver hierarchy

After determining the benchmark performance, subsets of the generated drivers are used for the ANN training to determine the driver relevance (Section 3.2). The ANNs trained with single input drivers at a time identify the relevance of the climatic variables as primary drivers of the ecosystem response. The driver yielding the highest performance of the modeled response will be called the dominant primary driver.

If the response is highly modulated by the dominant response (e.g. the response of photosynthesis by light), it acts like a carrier signal for the minor responses (e.g. temperature). In this case, the minor responses might be concealed and the network might not be able to pick up the underlying minor correlation directly. To overcome this problem the performance improvement ΔP_A, when adding a driver to the network mapping the dominant responses, is used as a measure of their relevance following a greedy search algorithm. The second greedy search steps determines the most relevant secondary driver, the third the most relevant tertiary driver.

The more greedy search steps, the larger the influence by cross-correlations among the input drivers. Therefore the greedy search algorithm should be stopped as early as possible. Usually three steps are enough to capture the concealed minor responses and additional steps do thus not provide new information about the driver relevances. The hierarchy of the remaining climatic drivers of the ecosystem response is obtained by ranking their performance improvement at the last step.

Since the greedy search algorithm only follows a local search, it might not always find the best subset. It is only sufficient if used in an ecophysiological context. For example, the photosynthetic photon flux density is known to be the primary driver of the carbon flux during daytime (see also results in Section 5.4).

4.5 Extracting the functional relationships

During the training, the very general initial ANN is constrained by the datasets. Afterwards, the ANN network function f (Section 3.3) represents the ecosystem response to its climatic controls as present in the data. This network function can be used to characterize the physiological properties of the ecosystem:

1. Its form (e.g. the basic shape, the offsets at the origin, or the saturation) shows the functional dependencies and can be used to derive the physiologically relevant parameters;

2. Its partial derivatives give information about the changes in the response with respect to changes in the input driver(s): Positive and negative

CHAPTER 4. PHASES OF THE METHODOLOGY

derivatives or potential turning points of the response may provide insight into the underlying processes;

3. Plotting the network function helps to visualize the functional dependencies on the climatic controls.

The hierarchy of the climatic controls allows this analysis to be confined to the relevant climatic drivers. First, ANNs trained only on the dominant primary driver are examined. The ANN models with only one dependent driver variable will be referred to as one-dimensional response models. Then, the ANN models trained simultaneously with the dominant climatic control plus secondary driver(s) are analyzed for their multivariate dependencies. The number of input drivers is expanded as long as there is a significant improvement in the ANN mapping performance with the following limitations:

Meaningful: The drivers should be physiologically meaningful.

Independent: The climatic drivers usually have obvious or hidden correlations that may distort the dependencies. Therefore, it is important to be aware of cross-dependencies and to keep the drivers as independent as possible.

Confounding: Confounding drivers should be included in the driver set in order to obtain robust relationships.

Minimal number: The degrees of freedom of the empirical model increase with each added driving variable. This may lead to a physiologically implausible mapping of the response. Consequently, the number of input drivers should be kept as low as possible.

4.6 Analyzing the sensitivities

The numerical partial derivatives characterize the changes in the ecosystem response with respect to changes in the climatic input driver(s) for a given data tuple (see Section 3.4). A large PaD means that a small change in this driver leads to a large change in the responding output, thus, a highly sensitive response. The sign of the derivative indicates whether the effect on the response is increasing (positive) or decreasing (negative).

A measure of the total sensitivity of the ecosystem response to a climatic driver over the range of analyzed data tuples is given by the absolute mean of the numerical partial derivatives and their positive and negative fractions. The higher the means, the more sensitive is the ecosystem response to a climatic driver.

4.7 Discussing the induced hypotheses

The results obtained are only meaningful if used in and put into the context of ecosystem physiology. Therefore, a detailed discussion of the induced hypotheses and their implications is an important phase of this methodology. In contrast to a hypothetic-deductive approach, the strength of a fully inductive approach to thrive only on the information present in the data is also its inherent limitation. In addition to the requirement of a high quality training dataset representative for the phenomena studied (see Phase 1 in Section 4.1), the following points have to be considered when interpreting the results:

Adequate response space: To reach the goal of modeling the overall response, an annual dataset is appropriate. If the interest is in the light response curve, the dataset should span time periods where the ecosystem stays in the same phenological and ecological states with respect to the photosynthesis response. For example, including months with leaves off would smear out the photosynthesis response. Taking summer months but including months with drought conditions might result in a light response curve where the saturation has a drop for the highest irradiances. This would look like photoinhibition, but is actually caused by the superposition of the light response curve with a reduced optimum NEP under water stress. As an alternative to limiting the dataset to the same state, the entire dataset can also be used but with an additional input variable describing the changing condition, for example a proxy for the water stress. With this, the ANN is able to distinguish between drought and non-drought conditions and map the responses accordingly.

Artifacts: To avoid modeling artifacts present in a specific dataset or non-obvious changes in the phenological or ecological states, the relationships identified should prove to be robust for different time periods, e.g., individual months vs. the whole summer period or summer months of different years.

Missing relevant driver: The ANN can show a good model performance, although a physiologically relevant driver was missing. This implies that the effect of the missing driver was mapped onto the included drivers through cross-correlations. In this case, the relationships found are usually neither independent nor robust: The mapped functional relationships will change as soon as another driver with some cross-correlation or the actual missing driver is added. If adding drivers does not change the main properties of the numerical partial derivatives, this is a good sign for robustness.

Confounding factors: The hidden biases or indirect effects caused by confounding phenological, ecological, or climatic factors are much harder to detect. To rule out known confounding factors, these can be added to the data used for training as observed or theoretical drivers. This way, their impact is included in the modeled response, provided that the confounding factors are not correlated with any of the other input drivers, that they are well defined over the whole range, and that they do not add too many degrees of freedom to the network. An alternative solution is to perform marginal sampling, where the dataset is grouped into subsets for certain ranges of the confounding factor. The ANN models are then trained on each of the subgroups. Robust relationships will hold true for all of the subgroups.

Ecophysiological plausibility: Since the ANN models are constrained solely by the data, some prior knowledge of ecosystem physiology is required to ensure a proper choice of the representative dataset and to judge the plausibility of the results under consideration of potential confounding factors. Only then does this inductive approach produce meaningful results.

The derived physiological properties are then compared to the existing hypotheses. This comparison may identify differences or validate hypotheses, but also has the potential to indicate new features present in the data. This last phase often leads to a refined response query, followed by another cycle through the seven phases.

All seven phases of the methodology are independent of a specific data domain. They can be generally applied to characterize an ecosystem's response to its climatic controls present in complex observational or synthetic datasets. The technical implementation of the phases was achieved through highly flexible setup for pattern generation and automated processing routines, as described in Appendix A. In the next part of the thesis, the different areas of application of the methodology are demonstrated for five examples.

Part II
Areas of application

Chapter 5
Characterizing ecosystem responses

The methodology developed in Part I can be used as a "vision panel" to ecosystem datasets for different areas of applications. The five areas presented in Part II range from characterizing the ecosystem response (Chapter 5), over testing a specific hypothesis (Chapter 6) and assessing competing light response curves (Chapter 7), to evaluating terrestrial biosphere models (Chapter 8). Additionally, the methodology can also be used to interpolate gaps in the flux measurements (Chapter 9). The applications are all demonstrated on the same ecosystem dataset: the carbon flux measurements at the Hainich forest.

In this chapter, the seven phases of the methodology are used to answer the question: What are the climatic controls of the daytime *NEP* fluxes during the active period of the Hainich forest *as present in the data*?

5.1 Specifying the response query: *Daytime NEP response*

The queried ecosystem response is the daytime response of the Hainich forest during the active period. For this analysis, quality-checked level-3 datasets of the three non-drought years (2000, 2001, and 2002) were obtained from the standardized Carboeurope IP database Papale *et al.* (2006). Since the interest is in the daytime response, only data with a photosynthetic photon flux density ($PPFD$) of more than 10 μmol photon m^{-2} s^{-1} was selected. Furthermore, the analysis was restricted to the active summer period with fully developed leaves from June to September and to best quality data (quality flag = 0) with complete input driver data. Additionally, five outlier data points of an exceptionally dry day, with a vapor pressure deficit (VPD) higher than 18 hPa, and twenty-seven data points with unrealistic values of the diffuse fractions ($f_{dif} = 0\%$ and $f_{dif} > 100\%$) were removed from the dataset. The total number of half-hourly data tuples analyzed was 3015.

The script used for generating the pattern for this response query is provided as an example pattern generation script in Appendix A.

5.2 Generating driver candidates

All observed driver data provided in the datasets were included as input driver candidates:

Rg	(Total) Global Radiation (W m^{-2})
Rd	Diffuse Global Radiation (W m^{-2})
$PPFD$	(Total) Photosynthetic Photon Flux Density (μmol photon m^{-2} s^{-1})
Rn	Net Radiation (W m^{-2})
Rr	Reflected Radiation (W m^{-2})
Rh	Relative Humidity (%)
SWC	Soil Water Content (%)
Ta	Air Temperature (°C)
Ts_1, Ts_2	Soil Temperature at 5 cm and at 30 cm Depths (°C)
Gs	Soil Heat Flux (W m^{-2})
$Precip$	Precipitation (mm)
ZL	Atmospheric Stability Parameter
WS	Wind Speed (m s^{-1})
$ustar$	Friction Velocity (m s^{-1})
WD	Wind Direction (°)

Several theoretical variables were generated, such as the following four latent variables inferred from the observations:

f_{dif} — Diffuse Fraction (0% - 100%) $= Rd/Rg$

$PPFD_{dif}$ — Diffuse $PPFD$ (μmol photon m^{-2} s^{-1}) $= f_{dif} \cdot PPFD$

$PPFD_{dir}$ — Direct $PPFD$ (μmol photon m^{-2} s^{-1}) $= PPFD - PPFD_{dif}$

VPD — Vapor Pressure Deficit (hPa) $= 6.1078 \cdot (1 - \frac{Rh}{100}) \cdot \exp(\frac{17.08085 \cdot Ta}{234.175 + Ta})$

Though latent variables do not increase the information content hidden in the dataset, they help improve the biological interpretability of the modeled ecosystem response. However, there are also other theoretical variables that might add *new* information such as:

R_{pot} Potential Radiation (Insolation) at the Top of Atmosphere (W m^{-2})
T_m Mean Daytime Air Temperature (°C)
Fuzzy   Fuzzy Variable for the Time of Day
NEP_{hh} *NEP* Measurement of the Previous Half-hour

Additionally, the measurement of the radiative temperature of the canopy was provided by the site PI Werner Kutsch:

Tc   Radiative Canopy Temperature (°C)

The set of twenty-five (observed and theoretical) variables comprised a broad driver candidate set for the characterization of the ecosystem response.

5.3 Benchmarking with all drivers

The ANN models trained with all climatic drivers yielded an R^2 of 94.4($\pm$0.1)%, where the value in brackets is the standard deviation over ten ANN training permutations. This benchmark means that 94.4% of the total variability of *NEP* in this half-hourly dataset can be explained from the twenty-five climatic drivers, see Figure 5.1, top. The projection onto *PPFD* in the bottom graph shows that almost the whole spread in the response is captured.

The standard deviation *SDev* of the model residuals can be used to get an estimate of the remaining error. The *SDev* of the benchmark ANN models binned to intervals of 5 μmol CO$_2$ m^{-2} s^{-1} varied between 1.2 and 2.7 μmol CO$_2$ m^{-2} s^{-1}, see Figure 5.2. To compare these results to previous work using paired observations or model residuals from Richardson *et al.* (2008), the relationship between *SDev* and the magnitude of *NEP* of the positive bins was determined using linear regression (dotted line):

$$SDev = 1.37(\pm 0.04) + 0.060(\pm 0.002) \cdot NEP. \quad (5.1)$$

The relationship obtained in equation 5.1 has an offset similar to the one

CHAPTER 5. CHARACTERIZING ECOSYSTEM RESPONSES

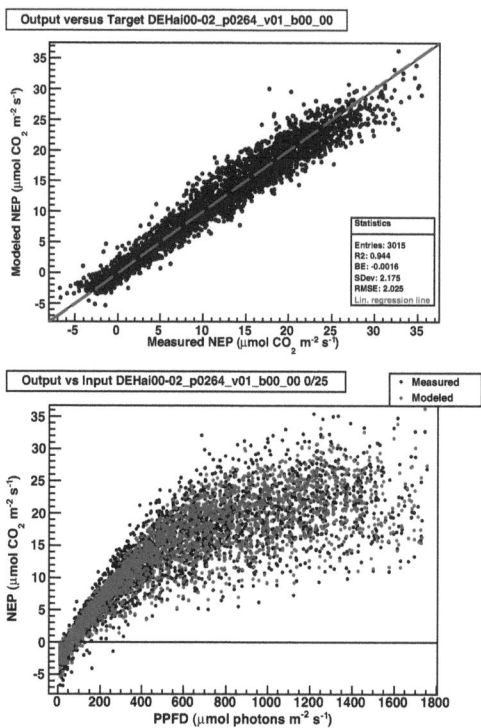

Figure 5.1: Benchmarking of the daytime *NEP* response of the Hainich forest with all twenty-five drivers. Top: Scatterplot of the modeled versus measured fluxes. Bottom: Projection of the modeled (red circles) and measured (black circles) fluxes onto *PPFD*.

previously reported for Hainich, but with only half the slope. This means that the random error estimated from the ANN benchmark models increases only half as fast with increasing flux magnitude. ANN training setups with different sets of input variables showed that the smaller increase can be attributed to including the diffuse radiation. A minimal configuration with only *PPFD*, $PPFD_{dif}$, *VPD*, and T_a as input drivers and only three to five nodes in the hidden layer of the ANN resulted already in almost half the slope. $PPFD_{dif}$ was not included in the analysis of Richardson et al. (2008).

CHAPTER 5. CHARACTERIZING ECOSYSTEM RESPONSES

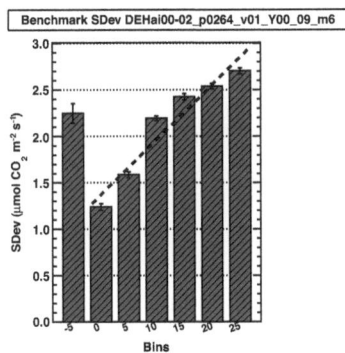

Figure 5.2: The standard deviation $SDev$ of the model residuals binned by the NEP flux magnitude in steps of 5 μmol CO_2 m^{-2} s^{-1}. (The dotted line is the linear regression of $SDev$ for the positive bins. The error bars show the standard deviation of ten ANN training permutations.)

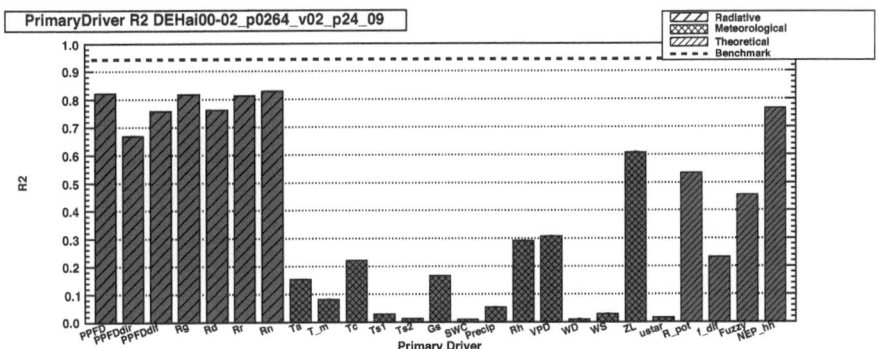

Figure 5.3: Primary R^2 performance of the ANN models trained with a single climatic driver at a time. (The dotted line is the benchmark performance of the ANN models trained with all twenty-five drivers. The error bars indicate the standard deviation of ten ANN training permutations; for most drivers this error bar is so small that it is not visible on the graph.)

5.4 Identifying the driver hierarchy

The ANN models trained with one climatic variable at a time showed the best performance with the radiative drivers, see Figure 5.3. There was only a slight

CHAPTER 5. CHARACTERIZING ECOSYSTEM RESPONSES

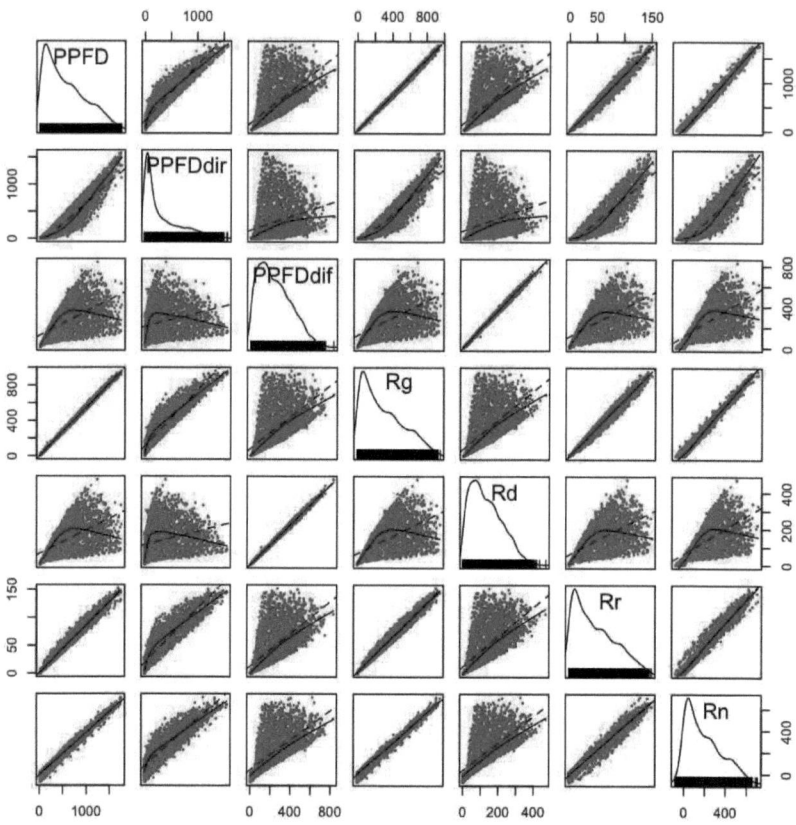

Figure 5.4: Scatterplot matrix of the radiative drivers. (Each column contains the same X axis and each row the same Y axis, with the variable name provided in the diagonal. The red dots are the scatter plot with a dashed line for the linear fit and a solid line for the lowess fit. The diagonal panel shows the sampling density and histogram of the variable.)

difference between the global radiation Rg ($R^2 = 81.65(\pm 0.01)\%$) and just the photosynthetically active radiation $PPFD$ ($R^2 = 82.16(\pm 0.01)\%$).

It is interesting to note that the primary R^2 performance of the reflected radiation Rr and the net radiation Rn was just as high. The scatterplot in

CHAPTER 5. CHARACTERIZING ECOSYSTEM RESPONSES

Figure 5.4 reveals, that the four drivers *PPFD*, *Rg*, *Rr*, and *Rn* have high linear correlations and, thus, almost the same information content. Based on common knowledge, it is understood that the response is driven by the incoming and not the reflected or net radiation. However, this example demonstrates how confounding variables may appear important that are not the actual cause of the effect. Since *PPFD* is biologically more relevant than *Rg*, it is used as *the* dominant driver for the following analysis of the secondary and tertiary drivers. This example stresses that the results of this purely empirical methodology have to be judged within the context of ecosystem physiology.

The direct part of the total radiation, $PPFD_{dir}$, was less relevant as a primary driver than the diffuse part $PPFD_{dif}$. This is probably due to the fact that the direct response saturates, whereas the diffuse response does not, as discussed below in Section 5.5.2. Another interesting aspect is that the *NEP* measurement for the previous half-hour explains 76.609($\pm$0.003)% of the total variability. This indicates the persistency of the meteorological conditions between successive half-hours and can be taken as a measure of the lower performance limit of models used for predictions.

The relevance of the other climatic controls as secondary controls was determined by training the ANNs with the dominating, biologically relevant *PPFD* plus one secondary climatic driver at a time, see Figure 5.5. The highest improvement in R^2 and *SDev* performance, thus the most relevant secondary control for the daytime *NEP* response, was the proportion of diffuse radiation. The performance improvement was the same whether this information was presented as $PPFD_{dir}$, $PPFD_{dif}$, R_d, or diffuse fraction f_{dif} to the total *PPFD* in the ANN models.

The ANN models with one of the diffuse proportion drivers added explained over 7% extra variability (Figure 5.5, top) and reduced the *SDev* by over twenty percent, from 3.7 to 2.9 μmol CO_2 m^{-2} s^{-1} (Figure 5.5, bottom). The R^2 of 89.5($\pm$0.1)% is close to the benchmarking performance, which indicates that the two drivers, total *PPFD* and proportion of diffuse light, explain almost all of the variability present in this half-hourly dataset.

Consequently, the tertiary drivers provided only marginal further improvement, see Figure 5.6. The remaining radiative drivers contained no additional information. The additional information was now contained in the meteorological conditions at the time of the measurement:

The canopy temperature *Tc* was slightly more relevant than the air temperature *Ta*, the heat soil flux *Gs* was more important than the two soil tempera-

CHAPTER 5. CHARACTERIZING ECOSYSTEM RESPONSES

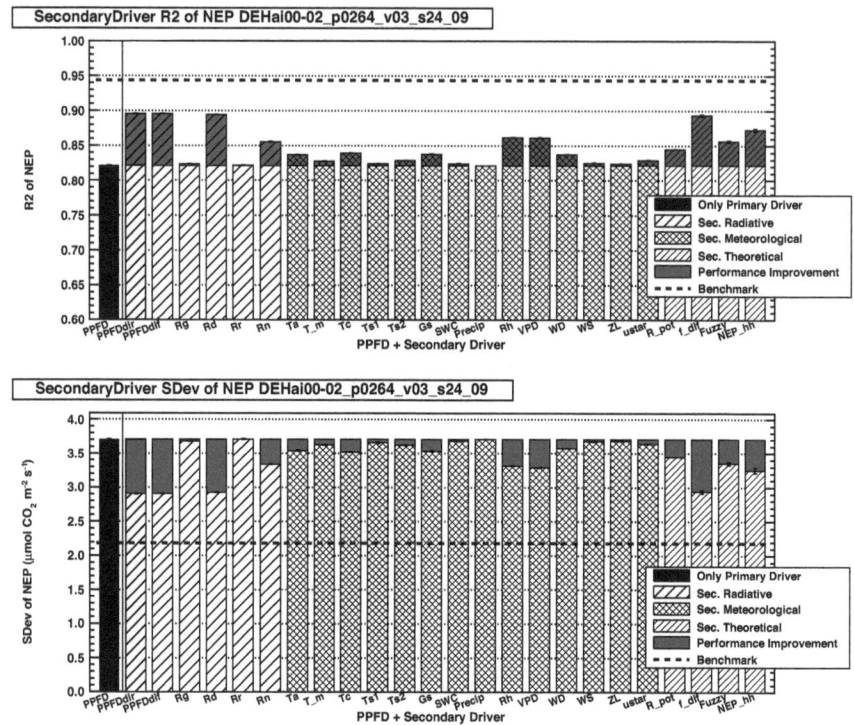

Figure 5.5: R^2 (top) and $SDev$ (bottom) performance of the ANN models trained with $PPFD$ plus a secondary climatic driver. The performance improvement (red) indicates the relevance as a secondary driver. (See Figure 5.3 for further explanations on the graphs.)

tures, Ts_1 and Ts_2. The precipitation $Precip$ was not relevant on the half-hourly scale, since rain does not contribute instantly to the NEP response. However, soil water content SWC was also of little relevance during these non-drought summers. Vapor pressure deficit VPD and relative humidity Rh both have the same significant relevance as a tertiary driver.

The parameters related to the eddy covariance technique, WD, Ws, ZL, and $ustar$, were all of little relevance, which is an indication of a clean dataset, i.e. these parameters did not induce systematic biases. The importance of the $Fuzzy$ variable for the time of day disappeared, when the confounding variable VPD (the "true cause") was included in the model runs. This can be attributed to the

CHAPTER 5. CHARACTERIZING ECOSYSTEM RESPONSES

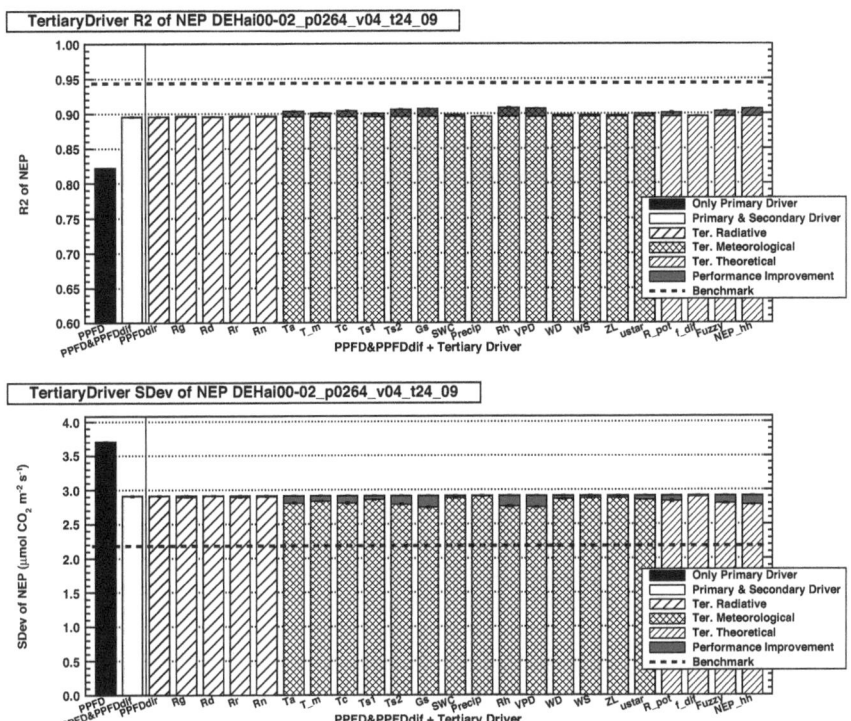

Figure 5.6: R^2 (top) and $SDev$ (bottom) performance of the ANN models trained with $PPFD$ and $PPFD_{dif}$ plus a tertiary climatic driver. The performance improvement (red) indicates the relevance as a tertiary driver. (See Figure 5.3 for further explanations on the graphs.)

fact that the *Fuzzy* variable was mapping the decreased *NEP* response caused by high *VPD* during the afternoon (not shown). The additional information in NEP_{hh} is of the same order of magnitude as in the meteorological drivers and can probably be attributed to similar meteorological conditions at the previous half-hour.

Using the mapping performance of the ANN models as a measure of the relevance of the climatic drivers allowed for a comprehensive analysis of the climatic controls of the daytime *NEP* response. Now that the relevant climatic controls have been identified, it is of interest to see what the mapped functional

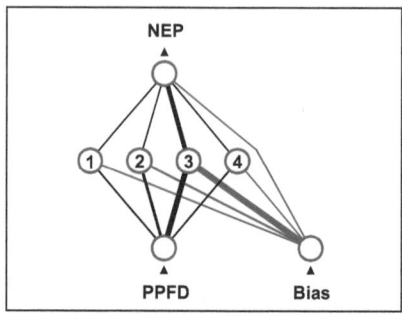

Figure 5.7: Final structure of one of the ANN models trained on the daytime *NEP* response with the single climatic driver *PPFD*. (The widths of the lines correspond to the magnitude of the weight parameters. The offset at each node is implemented as the weight to an additional bias node (red) with a constant input level of one.)

relationships actually looked like, as presented in the following sections.

5.5 Extracting the functional relationships

5.5.1 One-dimensional response to light

First the daytime *NEP* response to the dominating climatic driver, *PPFD*, was investigated. The final structure of one of the ANN models mapping *NEP(PFFD)* is shown in Figure 5.7. It had the following network function:

$$NEP(PPFD) = -38.4 + \frac{105.4}{\left(1 + 0.93 \cdot e^{\left(\frac{0.44}{1+14.6 \cdot e^{-0.0017 \cdot PPFD}} - \frac{0.29}{1+5.53 \cdot e^{-0.0017 \cdot PPFD}} - \frac{0.38}{1+5.08 \cdot e^{-0.0015 \cdot PPFD}} + \frac{2.60}{1+2.15 \cdot e^{0.0036 \cdot PPFD}}\right)}\right)} \quad (5.2)$$

Even though Equation 5.2 does not have biologically meaningful regression parameters, the physiological characteristics can be derived from the progression of this analytical function and its derivative, see Figure 5.8. The derivative starts off almost constant at the onset of light, corresponding closely to a linear initial slope. This initial slope of 0.050 μmol CO_2 m^{-2} s^{-1}/μmol photon m^{-2}

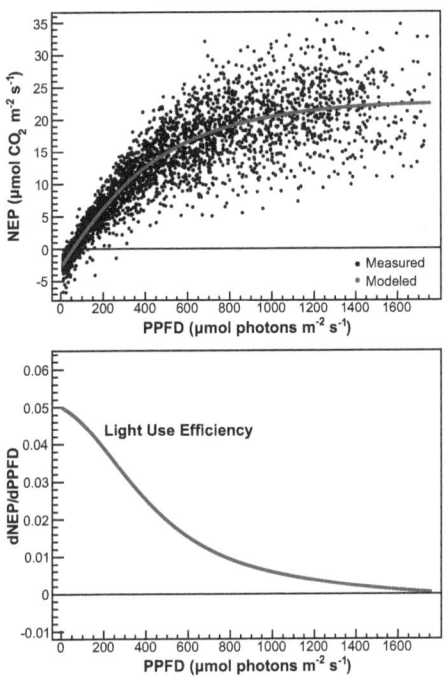

Figure 5.8: The daytime *NEP* response (top) and its numerical derivatives (bottom) modeled with *PFFD* as a single climatic driver.

s^{-1} is the maximum light use efficiency of the ecosystem, also called the initial quantum yield α. The offset of *NEP* at zero light is the daytime ecosystem respiration and has a value of -2.9 μmol CO_2 m^{-2} s^{-1}. Towards high *PPFD* values, the derivative approaches zero, denoting saturation of the *NEP* response. At the highest irradiance of 1750 μmol photon m^{-2} s^{-1}, the one-dimensional light response levels off to saturation (zero derivative) with an optimum *NEP* of 22.5 μmol CO_2 m^{-2} s^{-1}. The characteristics of the light response and the derived physiological parameters are discussed in more detail in Chapter 7.

The one-dimensional light response describes the mean behavior of the daytime *NEP* response with respect only to total *PPFD*. As shown in the last section on the hierarchy of the drivers, however, the ecosystem response is multi-dimensional, governed also by other climatic controls such as the proportion of diffuse light.

5.5.2 Multi-dimensional response functions

Before starting to look at the higher dimensional functional relationships, it is important to consider the correlations among the input drivers. The scatterplot in Figure 5.4 already showed the correlations between the radiative drivers. Most of them are related and thus correlated. However, $PPFD_{dir}$ and $PPFD_{dif}$ have little correlation and can be used to investigate the dependence of daytime NEP on the diffuse radiation.

Out of the meteorological variables, Ta, Tc, Ts_2, Gs, Rh, and VPD exhibited some relevance as tertiary drivers, see Figure 5.6 above. The scatterplots in Figure 5.9 shows that there is little correlation between these variables and $PPFD_{dif}$. The same is true for $PPFD_{dir}$ (not shown). However, the meteorological variables have some (cor-)relation amongst each other, especially Ta and VPD[1] or Ta, Tc, and Gs. Though there might be a confounding effect between VPD and $PPFD_{dif}$, there is no direct correlation of the two variables present in this dataset at the half-hourly scale.

As examples for multi-dimensional response functions, the relationship of $NEP(PPFD_{dir}, PPFD_{dif})$ and $NEP(PPFD_{dir}, PPFD_{dif}, VPD)$ are investigated in more detail below.

---

[1] Hence, if Ta is one of the network drivers, it is better to use the more independent Rh.

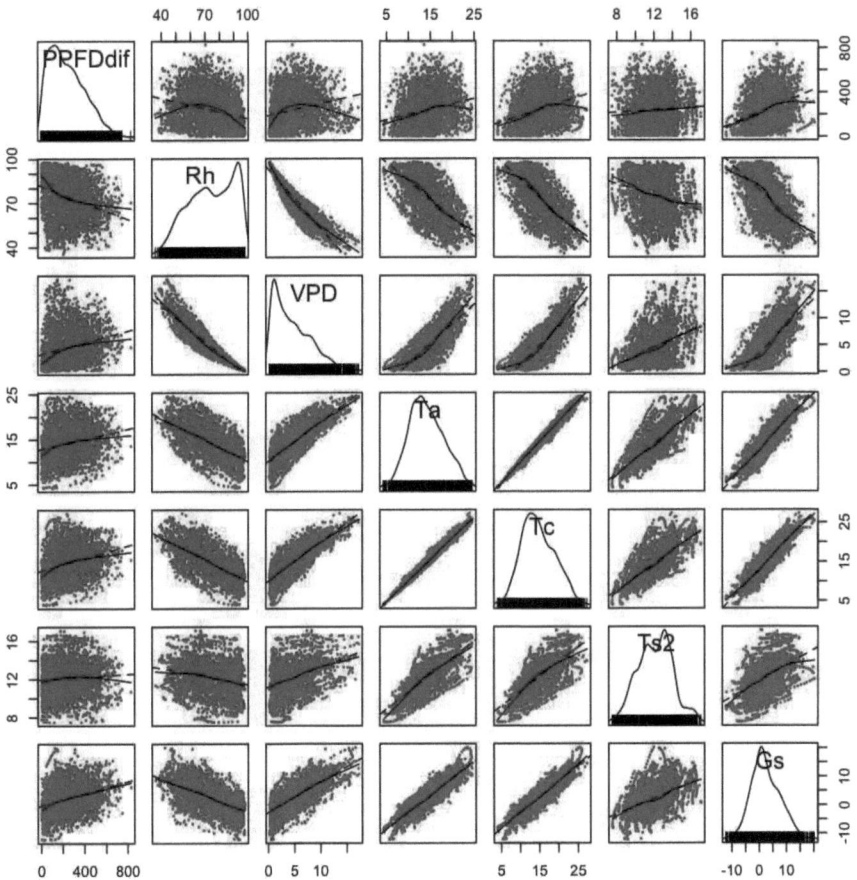

Figure 5.9: Scatterplot matrix of the meteorological drivers. (See Figure 5.4 for explanations on the graph.)

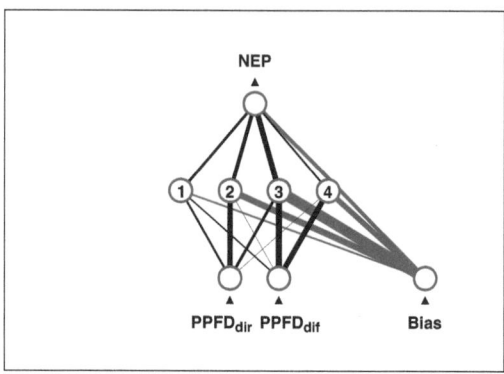

Figure 5.10: Final structure of one of the ANN models trained on the daytime *NEP* response with the two climatic drivers $PPFD_{dir}$ and $PPFD_{dif}$. (See Figure 5.7 for explanations on the graph.)

Two-dimensional response to direct and diffuse light

The proportion of diffuse radiation was the most relevant secondary control of the daytime *NEP* response at the Hainich forest (Section 5.4). Since total *PPFD* is (cor-)related with each of the four diffuse proportion drivers, the functional dependency of the *NEP* response on the diffuse proportion was extracted using direct *PPFD* and diffuse *PPFD* as input drivers.

The final structure of one of the ANN models of $NEP(PPFD_{dir}, PPFD_{dif})$ is shown in Figure 5.10. The projections onto the two climatic drivers in the top graph of Figure 5.11 show that the functional relationship of *NEP* to $PPFD_{dif}$ differs significantly from that to $PPFD_{dir}$. This fact is even more pronounced in the numerical partial derivatives in the bottom graphs. The initial quantum yield of $PPFD_{dif}$ is almost three times higher, its light use efficiency (magnitude of the derivative) is enhanced throughout the response, and the *NEP* response shows no saturation even for high $PPFD_{dif}$. These results are in full agreement with Gu *et al.* (2002), who found similarly enhanced light use efficiencies and weakened tendencies to cause canopy saturation for the diffuse radiation.

An even better grasp of the relationship of *NEP* to direct and diffuse *PPFD* is provided in the 3D-plot of the analytical network function in Figure 5.12. The

CHAPTER 5. CHARACTERIZING ECOSYSTEM RESPONSES

density of the data tuples (left graph) indicates that the *NEP* response is well constrained by the data. The simplicity of the ANN model (right graph) demonstrates that the extracted functional relationship $NEP(PPFD_{dir}, PPFD_{dif})$ is well suited to display and quantitatively characterize the response.

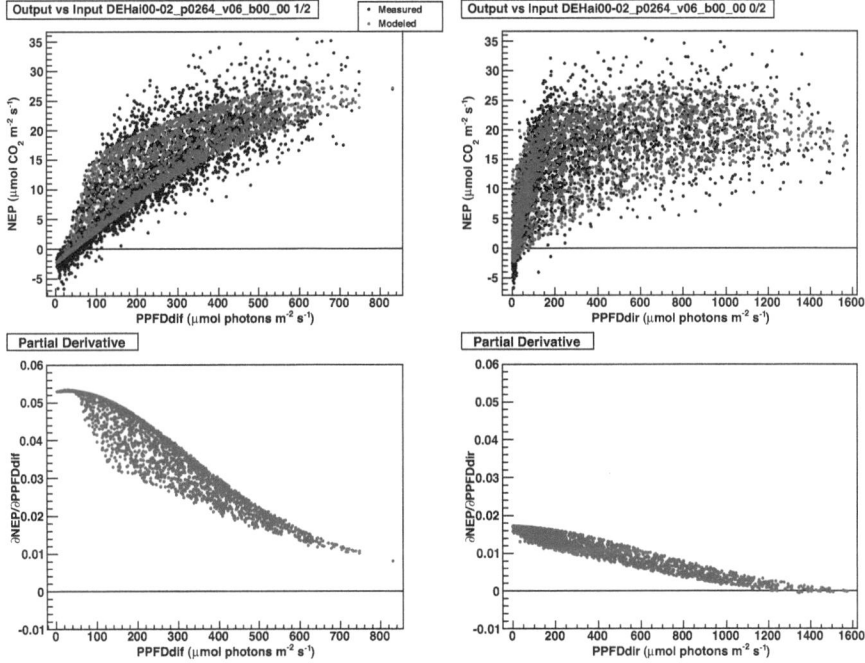

Figure 5.11: ANN model predictions (red circles) and half-hourly measurements (black circles) of the daytime *NEP* response plotted against the two climatic drivers diffuse $PPFD_{dif}$ (left) and direct $PPFD_{dir}$ (right). The graphs at the bottom show the numerical partial derivatives.

CHAPTER 5. CHARACTERIZING ECOSYSTEM RESPONSES

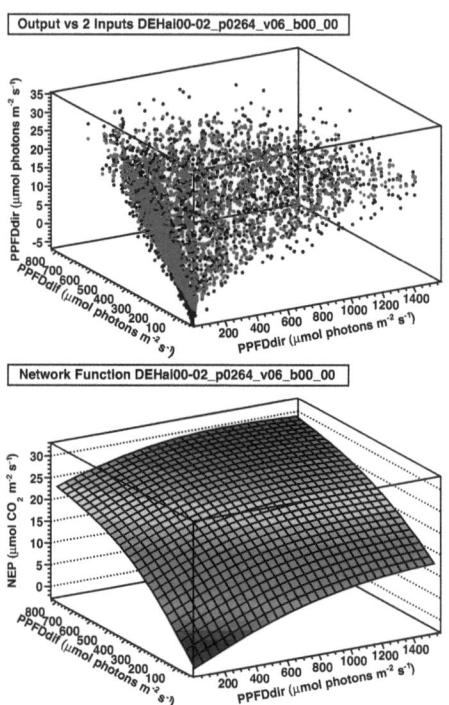

Figure 5.12: *Top:* 3D-plot of the daytime *NEP* response to the climatic controls diffuse $PPFD_{dif}$ and direct light $PPFD_{dir}$ for the individual half-hourly measurements, modeled (red) and measured (black). *Bottom:* Closed symbolic representation of the ANN model.

CHAPTER 5. CHARACTERIZING ECOSYSTEM RESPONSES

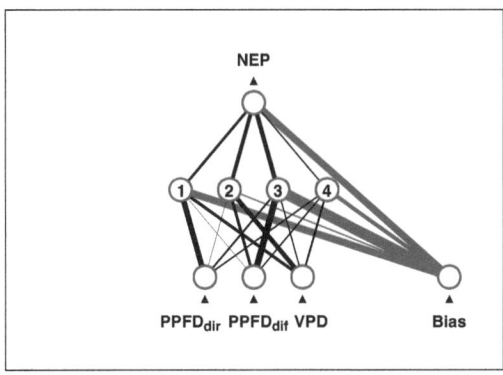

Figure 5.13: Final structure of one of the ANN models trained on the daytime *NEP* response with the three climatic drivers $PPFD_{dir}$, $PPFD_{dif}$ and *VPD*. (See Figure 5.7 for explanations on the graph.)

Additional effect of VPD

The daytime *NEP* response was also affected by the amount of air moisture. To investigate this effect, the daytime *NEP* response was modeled with $PPFD_{dir}$, $PPFD_{dif}$, and *VPD* as the climatic input drivers, see Figure 5.13. Adding *VPD* improved the R^2 by 1.2% to 90.7($\pm$0.1)% and reduced the *SDev* by 5% to 2.75 μmol CO_2 m^{-2} s^{-1}.

Since the daytime *NEP* response is now modeled with three inputs, the analytical function has too may dimensions to be directly visualized. For these multi-dimensional relationships, the numerical partial derivatives are of great value to examine their behavior (Figure 5.14 and Figure 5.15). The right bottom graph in Figure 5.15 shows that the daytime *NEP* response first exhibits a slight increase (positive derivative) with respect to *VPD*, then an optimum (zero derivative) around 4 hPa, and with increasing dryness of the air a strong down-regulating effect (negative derivative).

The functional relationships obtained for the daytime *NEP* response of the Hainich forest were very robust. Adding *VPD* did not change the behavior of

CHAPTER 5. CHARACTERIZING ECOSYSTEM RESPONSES

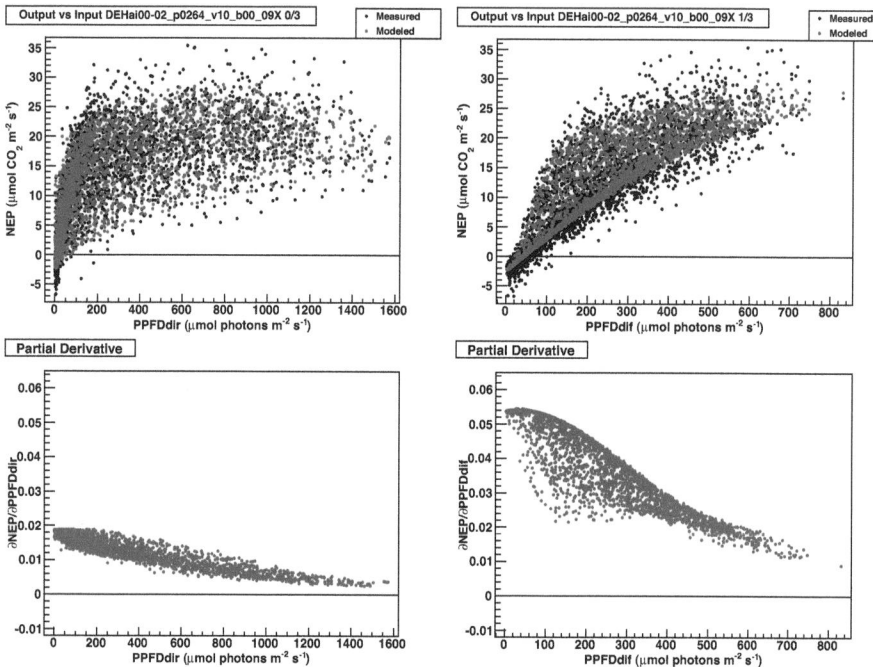

Figure 5.14: ANN model predictions of $NEP(PPFD_{dir}, PPFD_{dif}, VPD)$ (red circles) and measurements (black circles) plotted against the two of the three climatic drivers. (See next figure for third driver.) The graphs at the bottom show the numerical partial derivatives.

CHAPTER 5. CHARACTERIZING ECOSYSTEM RESPONSES

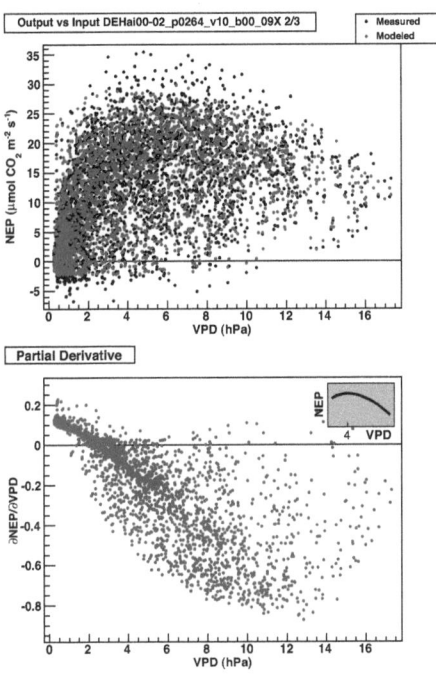

Figure 5.15: ANN model predictions of $NEP(PPFD_{dir}, PPFD_{dif}, VPD)$ (red circles) and measurements (black circles) plotted against the third climatic drivers. (See last figure for first two drivers.) The graphs at the bottom show the numerical partial derivatives. The small gray sketch depicts the functional relationship of NEP to VPD.

CHAPTER 5. CHARACTERIZING ECOSYSTEM RESPONSES

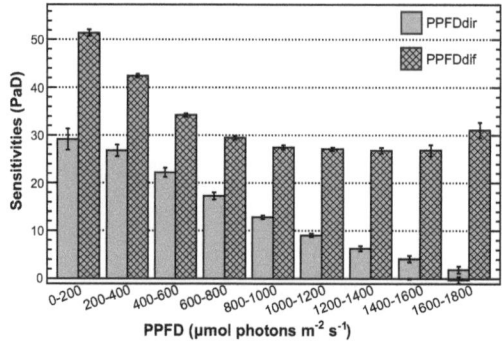

Figure 5.16: The ecosystem sensitivity of the daytime NEP response to direct $PPFD_{dir}$ and diffuse $PPFD_{dif}$, binned by total $PPFD$ in steps of 200 μmol photons m^{-2} s^{-1}. (The error bars show the standard deviation of ten ANN training permutations.)

the derivatives with respect to $PPFD_{dir}$ and $PPFD_{dif}$ in Figure 5.15, compared to Figure 5.11.[2]

5.6 Analyzing the sensitivities

The numerical partial derivatives supply information about the sensitivity of the ecosystem response to the climatic drivers. One interesting aspect is the change in the sensitivity of $NEP(PPFD_{dir}, PPFD_{dif})$ to diffuse and direct light over the range of the total available light. To examine this, the ecosystem dataset was binned (see Section 4.1) into subsets of total $PPFD$ for the analysis, see Figure 5.16. The enhanced light use efficiency of the diffuse radiation leads also to an enhanced sensitivity throughout the daytime NEP response, while the sensitivity to direct radiation decreases due to saturation. The effect of the enhanced sensitivity to $PPFD_{dif}$ is even more pronounced for high values of total $PPFD$, since the response does not saturate.

---

[2]This raises the question of whether the daytime NEP response to $PPFD_{dif}$ is really confounded by VPD at the half-hourly time scale. The two drivers also showed only little correlation in Figure 5.9. However, this argumentation cannot be turned around. If the diffuse radiation, which is the second most relevant driver, is missing as an input driver and only VPD is provided, then some of the effect that should actually be attributed to the diffuse radiation might be mapped on VPD.

CHAPTER 5. CHARACTERIZING ECOSYSTEM RESPONSES

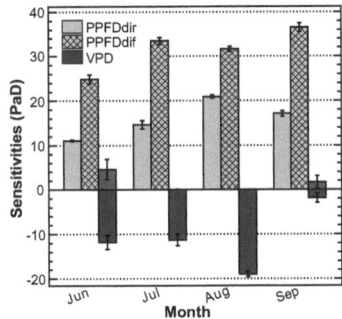

Figure 5.17: The positive and negative sensitivities of the daytime *NEP* response to $PPFD_{dir}$, $PPFD_{dif}$, and *VPD* during early afternoon hours, modeled separately for each month. (The error bars show the standard deviation of ten ANN training permutations.)

Grouping of the dataset into individual months for training and analysis can be used to investigate the monthly variability of $NEP(PPFD_{dir}, PPFD_{dif}, VPD)$ during the summer. To look primarily at the effect of *VPD*, only early afternoon hours (11:30-14:00) with stable light conditions but high changes in *VPD* were extracted for this analysis. Figure 5.17 shows that the negative sensitivity to *VPD* peaks in August, the hottest and driest month.

5.7 Discussing the induced hypotheses

Several hypotheses can be induced from the purely empirical modeling results above. The benchmark ANN models in Section 5.3 provided a measure of the total explainable variability in the dataset from the twenty-five generated climatic drivers.

The fact that the standard deviation of the ANN model residuals is even below the paired observation estimates from Richardson et al. (2008), corroborates the assumption that the remaining error and thus the unexplained variability can be mostly attributed to noise in the measurements. Moreover, it shows that the relevant climatic drivers were included in the training dataset, and that the ANN models were able to pick up the underlying correlations and fully capture the ecosystem response. A more comprehensive assessment of the reliability of the ANN models, such as their generalization ability and the impact of input uncertainty, can be found in Chapter 10.

The correlations mapped in the benchmark ANN models also permit the reconstruction of missing *NEP* measurements from the associated climatic data. Hence, these models can be used as a so called gap-filling technique as evaluated in Chapter 9.

During daytime, the *NEP* response in the active season is mainly dominated by photosynthesis and respiration plays only a minor role. Hence, the radiative variables were identified as the prevailing drivers of the response in Section 5.4. The behavior of the one-dimensional light response curve found in Section 5.5 and the magnitude of the derived physiological parameters agree well with the hypotheses on the light response in plant ecology for a deciduous broadleaf forest (Larcher, 2003). The agreement demonstrates that the inductive modeling framework is able to extract the underlying functional relationship directly from the data. Since this relationship was derived solely from the observations, without a priori assumptions, the agreement also provides an independent corroboration of the light response hypotheses. This independent corroboration is used in Chapter 7 to assess competing semi-empirical equations for the light response.

The new result, that the diffuse proportion of the radiation is the most relevant secondary driver of the daytime *NEP* response at Hainich during the active season, has long been suspected by site PI Alexander Knohl (personal communication, 2004, Knohl & Baldocchi, 2008). However, the methodology presented here provides a data-derived confirmation. The high input relevance and enhanced light use efficiency and sensitivity of $PPFD_{dif}$ compared to $PPFD_{dir}$ (Figures 5.5, 5.11, and Figure 5.16) stress the importance of the diffuse radiation for the ecosystem response.

To the author's knowledge, this is the first analytical representation and visualization of the relationship of the daytime response to both direct and diffuse light (Figures 5.11 and 5.12). Herein lies the strength of this inductive approach: in addition to the detection and quantification of the impact of diffuse radiation, it provides an explicit characterization of the functional relationship.

Unfortunately, the diffuse radiation is not measured at many of the flux tower sites, and there are no global datasets available. However, as the dominant secondary control of the half-hourly daytime *NEP* response, it should be included in ecosystem models trying to predict the carbon flux at half-hourly or hourly timescales. The hypotheses needed for the implementation can be based on the functional relationships derived by the ANN models from the data. The implementation of diffuse radiation can be expected to improve the agreement between the response functions extracted from observational data and the synthetic data of the two terrestrial biosphere models evaluated in Chapter 8.

The sensitivity analysis in Section 5.6 showed that the *NEP* response to *VPD* peaks in August, the hottest and driest month. In a study by Schulze (1970) on the carbon gas exchange of single beech trees in Sollingen, 100 km north-east of Hainich, the month of August also showed the strongest negative effect due to dry atmospheric conditions. Thus, the response to air moisture found at the tree level can be observed in the carbon flux measurements at the ecosystem level.

Overall, the methodology enabled a very comprehensive characterization of the daytime *NEP* response at the Hainich forest directly from the eddy covariance measurements. The FLUXNET database (www.fluxdata.org), with hundreds of site years available from flux towers all over the world, provides a huge study ground for further exploration. In the next chapter, rather than trying to characterize the ecosystem response, the methodology is used to set up a different kind of application: to test a specific hypothesis.

CHAPTER 5. CHARACTERIZING ECOSYSTEM RESPONSES

Chapter 6
Testing specific hypotheses

As a second area of application, the methodology can be used to test specific hypotheses by analyzing if a predicted effect is present in the ecosystem datasets. As an example, the net effect of the diffuse radiation at the Hainich forest is examined.

Diffuse radiation leads to an enhanced light use efficiency and thus an increased *NEP* response of the Hainich forest, as explored in previous chapter. However, less of the potential radiation R_{pot} is received at the surface for high diffuse fractions due to the absorption and reflection by clouds and aerosols, and less light leads to a decrease in the *NEP* response. It is still widely debated whether the overall effect is positive or negative. A detailed review of the opposing research results can be found in Knohl & Baldocchi (2008).

To get further insight into the net effect of diffuse radiation on *NEP*, Knohl & Baldocchi (2008) set up a biophysical multilayer model of the canopy for the Hainich site, tuned with the local measurements of the atmospheric transmission. Their model predicted an optimum in the *NEP* response for a diffuse fraction of 0.45. Whether this hypothesis can also be detected directly in the measurements is investigated in the following.

6.1 Specifying the response query: *Net effect of diffuse radiation*

Again, the ecosystem response queried is the daytime response of the Hainich forest during the active season. Therefore the same dataset as in the previous chapter was used, see Section 5.1 for a detailed description.

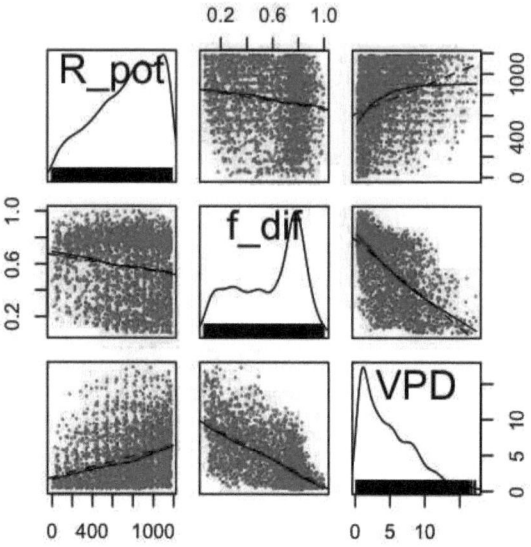

Figure 6.1: Scatterplot matrix of the three climatic variables potential radiation R_{pot}, diffuse fraction f_{dif}, and vapor pressure deficit VPD. (See Figure 5.4 for explanations on the graph.)

6.2 Generating driver candidates

The theoretical variable R_{pot} describes the total incoming radiation of the sun on top of the atmosphere that could potentially reach the ecosystem. The other variable of interest is the fraction of diffuse to total light f_{dif}. Since f_{dif} has some correlations with the vapor pressure deficit VPD, see Figure 6.1, and could therefore cause confounding effects, VPD was included in the analysis as a third input driver.

6.3 Benchmarking with all drivers

The structure of one of the ANN models trained with all three drivers R_{pot}, f_{dif}, and VPD can be found in Figure 6.2 (left). The ANN model captures 75.1($\pm$0.5)% of the variability in the half-hourly measurements, see Figure 6.2

CHAPTER 6. TESTING SPECIFIC HYPOTHESES

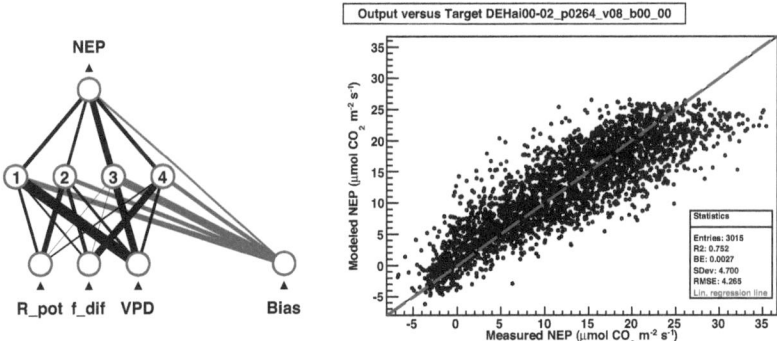

Figure 6.2: Network structure (left) and scatterplot of modeled versus measured NEP (right) for one of the ANN models trained with R_{pot}, f_{dif}, and VPD.

(right). Since R_{pot} is only an indirect driver of the daytime NEP response, the R^2 performance is lower than of the ANN models using one of the other radiative variables in Section 5.4.

6.4 Extracting the functional relationships

Since the daytime NEP response was again modeled with too many dimensions to directly visualize the analytical network function, the numerical partial derivatives are used to infer the functional relationships. The bottom graphs in Figure 6.3 and Figure 6.4 show the derivatives of the modeled $NEP(R_{pot}, f_{dif}, VPD)$ response with respect to each of the drivers. The derivative with respect to the potential radiation is, as expected, always positive. The derivative with respect to the vapor pressure deficit becomes clearly negative with increasing dryness of the air, as in Section 5.5.2.

The functional relationship of NEP to f_{dif} is depicted in the small gray sketch of Figure 6.4. The daytime NEP response is at first enhanced (positive derivative) until it reaches an optimum (zero derivative) and then reduced (negative derivative). By keeping the third driver VPD fixed, the two dimensional projection $NEP(R_{pot}, f_{dif})$ can be used to illustrate this behavior, see Figure 6.5. The data-derived net effect of the diffuse radiation hence also shows an optimum, which ranges from diffuse fractions of 28% to 44%.

The majority of the half-hours in the dataset are beyond the optimum range

CHAPTER 6. TESTING SPECIFIC HYPOTHESES

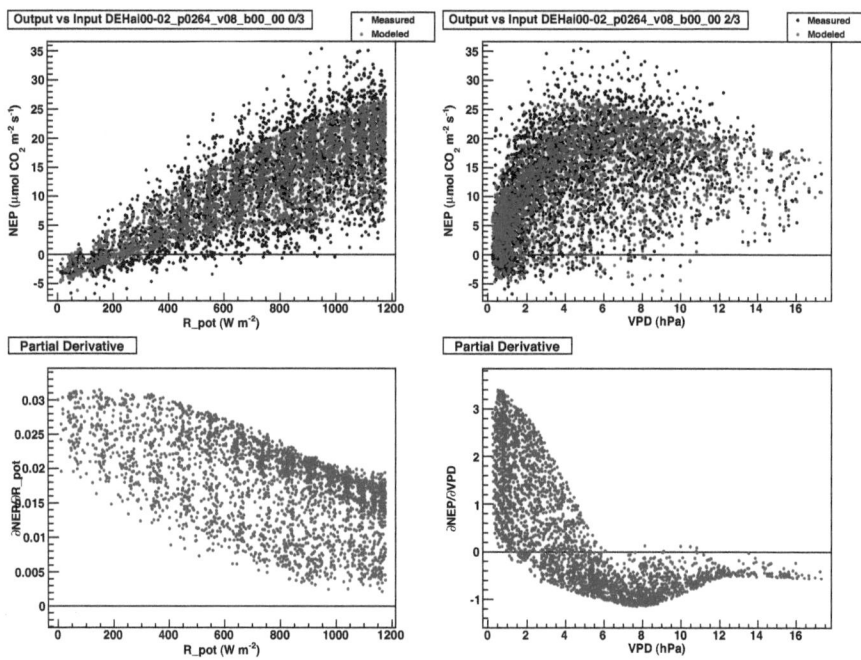

Figure 6.3: ANN model predictions of $NEP(R_{pot}, f_{dif}, VPD)$ (red circles) and measurements (black circles) plotted against R_{pot} and VPD.

of f_{dif}, as can be seen in Figure 6.4, middle. These counteract the positive effect on the *NEP* response for low diffuse fractions. Calculating the mean of all half-hourly numerical derivatives gives an indication of the overall effect. The average numerical derivative is -0.5 μmol CO_2 m^{-2} s^{-1} per half-hourly data point, thus a small but negative overall effect on *NEP* from diffuse radiation at the Hainich site for the daytime summer data of the years 2000 to 2002.

CHAPTER 6. TESTING SPECIFIC HYPOTHESES

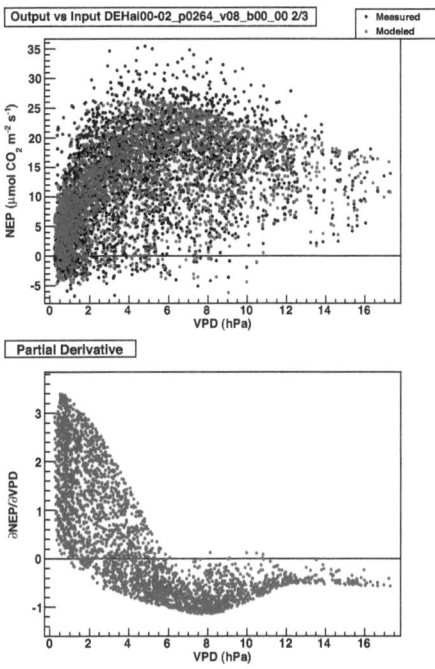

Figure 6.4: ANN model predictions of $NEP(R_{pot}, f_{dif}, VPD)$ (red circles) and measurements (black circles) plotted against f_{dif}. The small gray sketch depicts the functional relationship of NEP to f_{dif}.

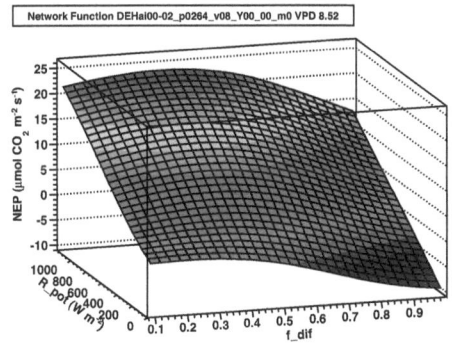

Figure 6.5: 3D-plot of the functional relationship of $NEP(R_{pot}, f_{dif})$ with the third input driver VPD fixed at the middle value (8.52 hPa).

6.5 Discussing the induced hypotheses

The data-derived net effect of the diffuse radiation revealed an optimum range of 28% to 44%, which is close to the optimum of 45% predicted by (Knohl & Baldocchi, 2008) using the biophysical multilayer model.

Both optima are at diffuse fractions, where there is less diffuse than direct light. Since the light conditions at the Hainich forest were mostly beyond the optimum diffuse fraction, the overall effect of the diffuse light was on average slightly negative for the three years 2000 to 2002. Since these are very site specific results, it would be of interest to analyze how this response is manifested in other types of ecosystems.

The above application example showed how the methodology can be used for testing new hypotheses. Rather than setting up a complex model, this "vision panel" to the data might be able to provide direct and instant answers. The data-derived answers can also be used to assess competing hypotheses, as demonstrated in the next chapter.

Chapter 7
Assessing competing semi-empirical equations

Often, ecophysiological processes such as the response to light are implemented in models as hypothesis-based "semi-empirical" equations. Usually, the assessment of these equations is based on their fit performance and their robustness for use in non-linear regression (NLR). These two criteria alone, however, are not sufficient if these equations are also used to derive meaningful ecophysiological parameters. Here, the presented methodology can be used to look at the functional relationships present in the ecosystem datasets. The purely empirical relationships thus found can then serve as data-derived references to assess competing working hypotheses.

The application of the methodology as an assessment tool is demonstrated for light response curves, which are one-dimensional semi-empirical equations for $NEP(PPFD)$ used to model the daytime NEP response to light. They are widely applied to carbon flux datasets for:

- Identification of the initial quantum yield and the optimum gross primary production of an ecosystem (Zhang *et al.*, 2006),

- Partitioning of the net carbon fluxes into respiration and gross primary production (Gilmanov *et al.*, 2003),

- Gap-filling of missing carbon flux data points (Falge *et al.*, 2001; Moffat *et al.*, 2007),

- Estimation of nighttime respiration from daytime measurements (Lasslop *et al.*, 2009), or

CHAPTER 7. ASSESSING COMPETING SEMI-EMPIRICAL EQUATIONS

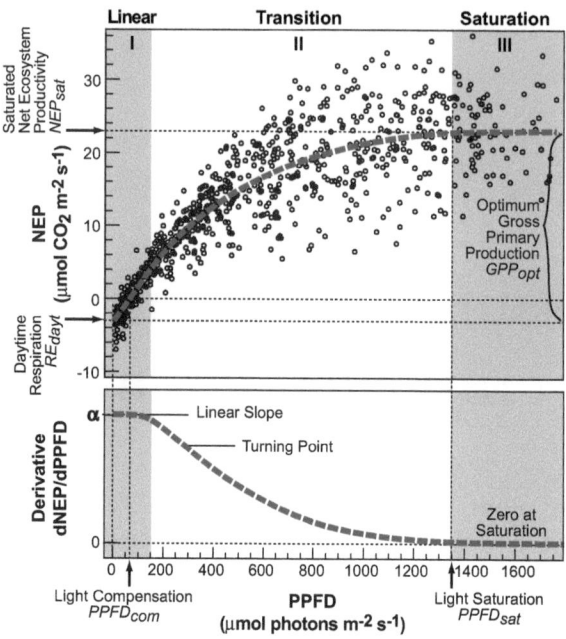

Figure 7.1: Sketch of a typical light response curve $NEP(PPFD)$ (red dashed line, top) and its derivative (bottom). The black circles are the half-hourly measurements of the daytime NEP response of the Hainich forest.

- Parameterization of the light response in higher complexity models such as TBMs (e.g. Knorr & Kattge, 2005; Krinner et al., 2005).

7.1 Specifying the response query: *Light response curve*

Since this application of the methodology aims to assess competing mathematical representations, the established theory of the light response (e.g. Larcher, 2003) is reviewed and the key properties formulated as mathematical constraints in the following.

The sketch in Figure 7.1 illustrates the light response. The measured net ecosystem productivity NEP is equal to the gross primary production GPP (up-

CHAPTER 7. ASSESSING COMPETING SEMI-EMPIRICAL EQUATIONS

take of carbon by photosynthesis) minus the ecosystem respiration ER (release of carbon):

$$NEP = GPP - ER. \tag{7.1}$$

At the light intercept to zero radiation, there is no uptake but only the daytime respiration ER_{dayt}:

$$NEP(PPFD) = -ER_{dayt} \text{ for } PPFD = 0. \tag{7.2}$$

For low irradiance, the uptake is very small and respiration exceeds the CO_2 uptake by photosynthesis until the light compensation point $PPFD_{com}$. Here the carbon release and the uptake are at equilibrium:

$$NEP(PPFD_{com}) = 0. \tag{7.3}$$

At very high irradiance, $PPFD_{sat}$, the photosynthetic process saturates:

$$NEP(PPFD_{sat}) = const = NEP_{sat}. \tag{7.4}$$

The optimum gross primary production GPP_{opt} is the light-saturated (maximum) photosynthesis rate:

$$GPP_{opt} = NEP_{sat} + ER_{dayt}. \tag{7.5}$$

The characteristics of the light response curve can be divided into three phases: (1) linear increase, (2) transition, and (3) saturation. In each phase, the curve has distinctive basic properties in its shape (Figure 7.1, top), which can be best depicted in the slope (first derivative, Figure 7.1, bottom).

Phase 1: At first the CO_2 uptake is linear, driven by a light-limited photochemical process. The proportionality factor (slope) of the maximum efficiency of light utilization is the initial apparent quantum yield α:

$$\frac{dNEP(PPFD)}{dPPFD} = \alpha \text{ for small } PPFD. \tag{7.6}$$

A linear response means, that the function has a constant slope and the first derivative thus intercepts the NEP-axis horizontally towards zero light.[1]

---

[1] Below the light compensation point $PPFD_{com}$, the Kok effect may cause a break from linearity (Kok, 1948; Atkin et al., 2000); this effect is not further considered here.

Phase 2: In the transition phase, enzymatic and carbon supply processes start to limit the photosynthetic uptake of carbon. The response levels off from a steep linear increase towards saturation. This change implies a turning point $PPFD_{turn}$ in the slope (first derivative) from concave-down to concave up. At $PPFD_{turn}$, the curvature (second derivative) is extremal and the third derivative thus zero:

$$\frac{d^3 NEP(PPFD_{turn})}{dPPFD^3} = 0. \tag{7.7}$$

Phase 3: In the saturation phase, the light response for C3 plants fully saturates. The response becomes flat and its slope zero:

$$\frac{dNEP(PPFD)}{dPPFD} = 0 \text{ for } PPFD > PPFD_{sat}. \tag{7.8}$$

The estimated physiological parameters depend on the mathematical characteristics of the light response curve in these three phases. To investigate the one-dimensional response to light (see also Section 5.5.1), the daytime NEP response during the active season is of interest. To avoid effects of phenology or heat stress later in the summer at the Hainich forest, the data selection criteria described in Chapter 5 were constrained even further to data from spring and early summer of the two years 2000 and 2001. Furthermore, only days with similar mean daytime temperature (i.e. between 12.5°C and 17.5°C) were chosen to reduce the effect of temperature. This resulted in a dataset that comprised approximately 700 data points of best quality data from approximately 30 separate days.

7.2 Benchmarking with all drivers

Again, benchmarking can be used to determine how much of the variability in the NEP fluxes can be explained. The ANN models for $NEP(PPFD)$ had an R^2 performance of 79.00 (± 0.01)%. The standard deviation $SDev$ of the ANN model residuals can be seen in Figure 7.2. To compare these residuals with the previous benchmarking result with 25 climatic drivers in Section 5.3, the linear relationship of the model residuals and the magnitude of NEP was calculated for positive fluxes:

$$SDev = 2.5(\pm 0.3) + 0.11(\pm 0.02) \cdot NEP. \tag{7.9}$$

CHAPTER 7. ASSESSING COMPETING SEMI-EMPIRICAL EQUATIONS

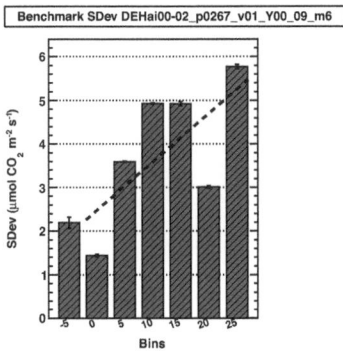

Figure 7.2: The standard deviation *SDev* of the ANN model residuals binned by the *NEP* flux magnitude in steps of 5 μmol CO_2 m^{-2} s^{-1}. (See Figure 5.2 for explanations on the graph.)

Though the dataset had been limited to reduce the effects of phenology and temperature, *PPFD* alone is not sufficient to capture the full *NEP* response. Equation 7.9 has twice the offset and twice the slope compared to Equation 5.1 of the benchmark network trained with all twenty-five climatic drivers.

An extra 10% of this unexplained variability can be attributed to the diffuse light proportion, which was identified as the second most important driver in Section 5.4. Modeling of the daytime *NEP* response to total *PPFD* as a one-dimensional function can only be expected to resemble the mean behavior to light.

7.3 Extracting the functional relationships

ANNs trained with the backpropagation algorithm can be regarded as nonparametric nonlinear regressions. After training, the *NEP* response to *PPFD* was mapped by the ANN model, see Figure 7.3. The data-derived *NEP* response to *PPFD* exhibits the expected behavior: a steep, almost linear initial increase leveling off to saturation for high *PPFD*.

CHAPTER 7. ASSESSING COMPETING SEMI-EMPIRICAL EQUATIONS

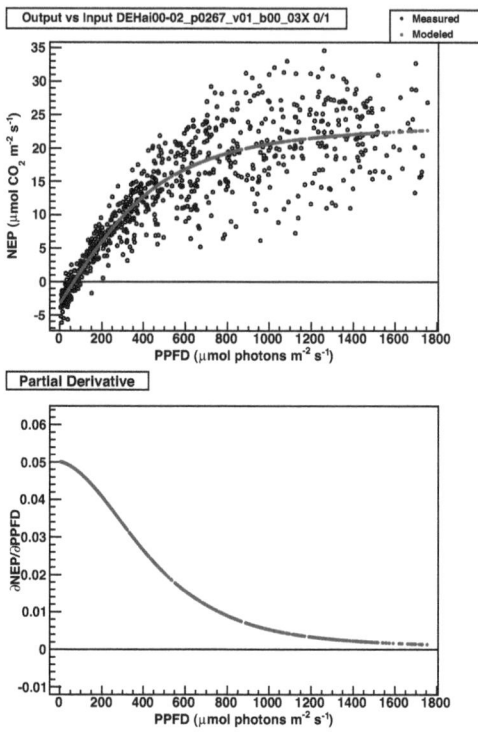

Figure 7.3: The ANN light response curve (top) and its numerical derivatives (bottom). The black circles are the half-hourly measurements of the daytime *NEP* response of the Hainich forest.

The analytical network function of the ANN model in Figure 7.3 was:

$$NEP(PPFD) = -36.73 + \frac{101.91}{\left(1 + e^{-0.302 \left| \frac{0.461}{1+e^{1.197+0.829 \cdot (-1.005+0.00114 \cdot PPFD)}} + \frac{2.627}{1+e^{3.747+3.110 \cdot (-1.005+0.00114 \cdot PPFD)}} \right.}\right)} \quad (7.10)$$

Though the ANN weight parameters have no direct physiological meaning, the mathematical characteristics of this network function can be used to derive the physiological parameters using Equations 7.2 to 7.8 above:

Phase 1: At low *PPFD*, the light response is almost linear, with a horizontal slope. The light compensation point $PPFD_{com}$ is at 66 μmol photon m^{-2} s^{-1}. The intercept with zero gives an initial quantum yield α of 0.0501 and a daytime respiration ER_{dayt} of 3.26 μmol CO$_2$ m^{-2} s^{-1}.

Phase 2: The slope then curves down to zero with a turning point, $PPFD_{turn}$, at 276 μmol photon m^{-2} s^{-1}.

Phase 3: For $PPFD$ higher than 1349[1]μmol photon m^{-2} s^{-1}, the response saturates and the derivative approaches zero. The optimum gross primary production GPP_{opt} is 25.95 μmol CO$_2$ m^{-2} s^{-1}.

7.4 Discussing the induced hypotheses

The response of plants to light is dominated by photosynthesis. This assimilation of carbon using light has been widely studied since Wolkoff (1866) and there exists a well established theory at the leaf and plant level. The light response curve and its derivative, mapped by the ANN model, display the behavior expected from plant ecology for a deciduous broadleaf forest (Larcher, 2003). With the eddy covariance technique, the measurements have been brought to the ecosystem level. The fact that the ANN model was able to detect this response directly in the measurements (without prior assumptions about the functional shape), independently confirms that the theory derived on the plant level can be scaled to the ecosystem level.

In the following, all six light response curves used by the eddy covariance community plus a new suggestion are examined. These semi-empirical equations have a similar curve progression in the mean fit region and their fit performance is thus very similar. However, there are large discrepancies at the edges, where the physiological parameters are derived. This had been noticed by some authors, as the parameters became implausible, and led to personal preferences of certain equations (e.g. Aubinet et al., 2001; Gilmanov et al., 2003). However, an objective analysis of the characteristics of the light response curves and their ecosystem physiological appropriateness remained open and is now attempted.

---

[1] The value of $PPFD_{sat}$ was calculated at the point where the derivative of the response decreases below 0.0025 ($\sim 1/20\alpha$).

CHAPTER 7. ASSESSING COMPETING SEMI-EMPIRICAL EQUATIONS

Evaluation of the light response curves

The non-linear regression of the seven semi-empirical light response curves was performed on the same ecosystem dataset of the Hainich forest as used for the ANNs described above. All model results can be found in Figure 7.4. The most crucial areas of the light response are right at the edges, since these determine the magnitude of α, ER_{dayt}, and GPP_{opt}. It is precisely here, however, that the characteristics of the prescribed functional relationships of some of the semi-empirical functions differ from both, the features expected from basic plant physiology and the features present in the observational data (as mapped independently with the ANN models). This led to large differences in the estimated physiological parameters, despite similar fit performances, see Table 7.1.

The seven equations are evaluated in the following based not only on model fit performance but also on their mathematical properties. An appropriate mathematical description of the ecosystem response to light should correctly reflect the three phases in order to provide an adequate estimate of the physiological parameters. The purely empirical estimates of the physiological parameters from the ANN model serve as the reference.

7.4.1 Equation 1: Linear function with upper limit

The most basic mathematical form of the light response is a piecewise linear function terminated by a stationary upper limit (Blackman, 1905):

$$NEP(PPFD) = \alpha \cdot PPFD - ER_{dayt} \quad \text{for } PPFD < PPFD_{sat,turn}, \quad (7.11)$$
$$NEP(PPFD) = GPP_{opt} - ER_{dayt} \quad \text{for } PPFD \geq PPFD_{sat,turn}. \quad (7.12)$$

Evaluation: These equations do not capture the overall response (Figure 7.4), but only correctly resemble the first and third phases. However, the linear fit performed between 50 and 150 μmol photon m^{-2} s^{-1} and the limit fit performed above 1300 μmol photon m^{-2} s^{-1} provided good estimates for α, ER_{dayt}, and GPP_{opt} (Table 7.1). Though this function obviously oversimplifies the light response of photosynthesis and is not suitable for modeling the overall NEP response, it is capable of approximating the three physiological parameters.

CHAPTER 7. ASSESSING COMPETING SEMI-EMPIRICAL EQUATIONS

	Artificial Neural Netw.	Linear Func. with Limit	Rect. Hyperbola	Mod. Rect. Hyperbola	Non-rect. Hyperbola	Smith Sigmoid	Log. Sigmoid	Exp. Saturation
R^2	79.0%	67.8%	78.7%	78.7%	79.0%	79.0%	79.0%	79.0%
$RSME$	4.16	5.15	4.19	4.19	4.16	4.16	4.16	4.15
Quantum yield α	0.0501	0.0518	0.0846	0.0846	0.0530	0.0493	0.0450	0.0639
	—	(4%)	**(69%)**	**(69%)**	(6%)	(-2%)	(-10%)	**(27%)**
ER_{dayt}	3.26	3.36	4.72	4.72	3.42	3.28	2.88	3.91
	—	(3%)	**(45%)**	**(45%)**	(5%)	(1%)	(-12%)	**(20%)**
GPP_{opt}	25.95	25.65	35.67	29.45	29.02	27.4	25.16	27.2
	—	(-1%)	**(37%)**	(14%)	(12%)	(6%)	(-3%)	(5%)
$PPFD_{com}$	66	65	64	64	67	67	64	66
$PPFD_{sat}$	1349	495	>2000	>2000	1466	1395	1189	1381
$PPFD_{turn}$	276	495	none	none	274	277	368	2000
Curvature d	—	—	—	—	0.75	—	—	—

Table 7.1: Fit performance and estimates of the physiological parameters for modeling the light response at Hainich. (Stated in brackets is the deviation from the estimates of the artificial neural network, with percentages larger than 20% marked in bold.)

7.4.2 Equation 2: Rectangular hyperbola

A hyperbola is the section of a conical cylinder. Three different mathematical representations are commonly used for modeling the light response: the rectangular hyperbola, the modified rectangular hyperbola, and the non-rectangular hyperbola.

Michaelis & Menten (1913) originally derived a rectangular hyperbola for enzyme reaction rates as a function of the substrate concentration. Later research on the kinetics of photosynthesis also yielded a rectangular hyperbola (e.g. Baly, 1935; Smith, 1938; Rabinowitch, 1951):

$$NEP(PPFD)) = \frac{\alpha \cdot PPFD \cdot GPP_{opt}}{GPP_{opt} + \alpha \cdot PPFD} - ER_{dayt}. \tag{7.13}$$

Evaluation: The behavior of this function is contrary to the required characteristics of the light response, especially at the edges (Figure 7.4). At low light, instead of leveling off to a constant slope, the rectangular hyperbola's

CHAPTER 7. ASSESSING COMPETING SEMI-EMPIRICAL EQUATIONS

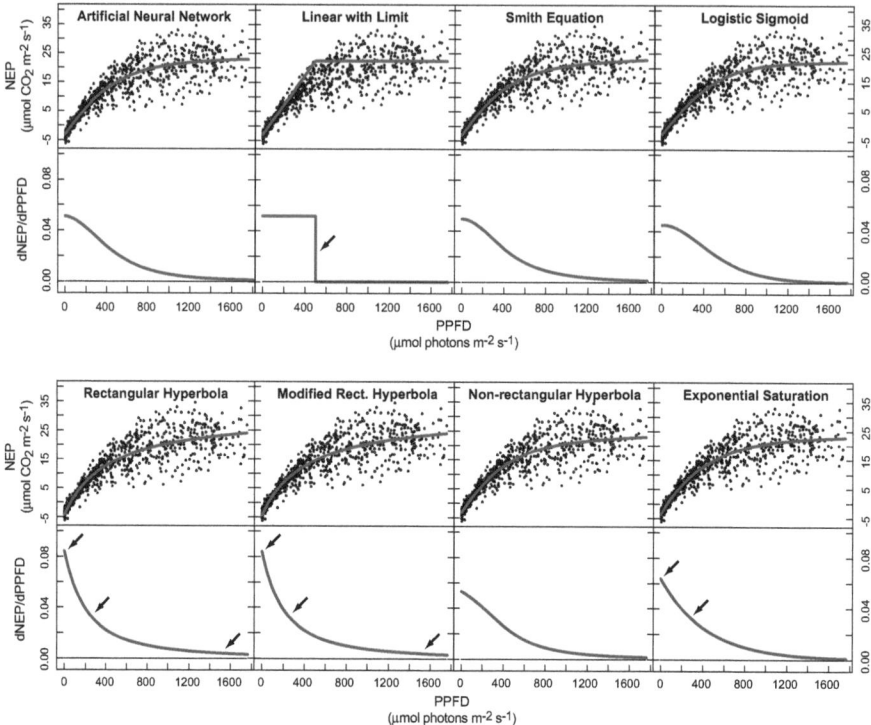

Figure 7.4: The ANN and the seven semi-empirical light response curves fitted to the daytime NEP measurements of the Hainich forest (black circles). Plotted below are the derivatives with incorrect behavior indicated by black arrows.

slope peaks to its highest value. This led to major overestimates in α of 69% and in ER_{dayt} of 45% (Table 7.1). At high irradiance, the saturation is hyperbolic and only fully saturates at infinity. Therefore, the derivative of the hyperbola shows no evidence of having reached saturation, even for the highest observed $PPFD$. This resulted in overestimates in GPP_{opt} of up to 37%

7.4.3 Equation 3: Modified rectangular hyperbola

The modified rectangular hyperbola is the same function as Equation 7.13, but with the saturation parameter GPP_{opt} determined not at infinity but at a $PPFD$ of 2000 μmol photon m^{-2} s^{-1} (Falge et al., 2001):

$$NEP(PPFD) = \frac{\alpha \cdot PPFD}{1 - (PPFD/2000) + (\alpha \cdot PPFD/GPP_{opt})} - ER_{dayt}. \quad (7.14)$$

Evaluation: Since the saturation parameter GPP_{opt} is determined not at infinity, but at $PPFD$ of 2000 μmol photon m^{-2} s^{-1}, the overestimate of GPP_{opt} was reduced to 14% (Table 7.1).

7.4.4 Equation 4: Non-rectangular hyperbola

The non-rectangular hyperbola is a more general hyperbola derived for photo-stationary concentrations (Rabinowitch, 1951):

$$NEP(PPFD) = \frac{\alpha \cdot PPFD + GPP_{opt} - \sqrt{(\alpha \cdot PPFD + GPP_{opt})^2 - 4 \cdot \alpha \cdot GPP_{opt} \cdot d \cdot PPFD}}{2d} - ER_{dayt}. \quad (7.15)$$

The additional curvature parameter $d \in [0, 1]$ changes the shape of this hyperbola from a rectangular hyperbola for $d \to 0$ to a linear response function with an upper limit for $d \to 1$.

Evaluation: The freedom of the extra parameter d cannot be fully constrained due to the variability in the eddy covariance measurements. To demonstrate this, the non-rectangular hyperbola was fitted to the dataset for fixed values of d, see Figure 7.5. For d between 0.1 and 0.9, the curvature of the non-rectangular hyperbola changes only slightly and all graphs are well within the measurement noise. This led to an equal R^2 performance (78.9±0.1%), whereas the initial derivative, the quantum yield α, varied by almost 100%. At the cell and plant levels, the measurements are more confined and the curvature is clearly present in the data. There d typically ranges from 0.7 to 0.99 and can be related to changes in the biochemistry (Ögren, 1993).

However, constraining the curvature parameter d is critical, since the non-rectangular hyperbola only fulfills the criteria of the light response at high values of d. With a high d of 0.75, the non-linear regression on this dataset yielded good estimates of α, ER_{dayt}, and GPP_{opt} (Table 7.1).

CHAPTER 7. ASSESSING COMPETING SEMI-EMPIRICAL EQUATIONS

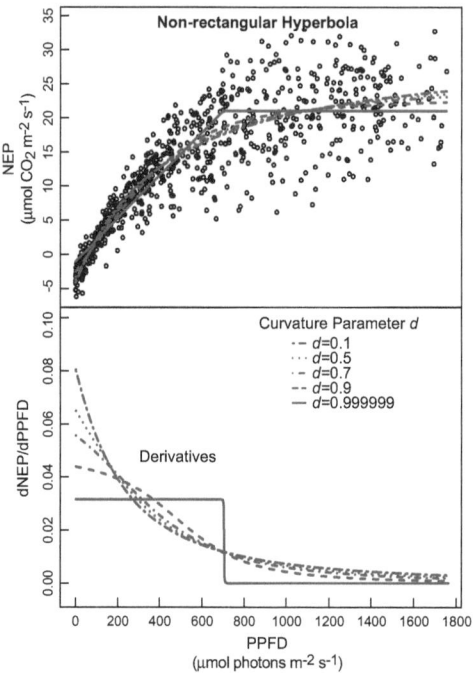

Figure 7.5: The non-rectangular hyperbola (top) and its derivative (bottom), fitted with fixed values of the curvature parameter d.

7.4.5 Equation 5: Smith sigmoid

Another class of functions used to describe the light response are sigmoids with an S-curve shape: the Smith sigmoid and the logistic sigmoid.

The Smith equation has the following algebraic function (Smith, 1937, 1938):

$$NEP(PPFD) = \frac{\alpha \cdot GPP_{opt} \cdot PPFD}{\sqrt{GPP_{opt}^2 + (\alpha \cdot PPFD)^2}} - ER_{dayt}. \qquad (7.16)$$

Evaluation: The Smith equation is a sigmoid function that fulfills the characteristics of the light response in all three phases (Figure 7.4). Therefore, the estimated physiological parameters agreed well with estimates from the ANN model for all three sites (Table 7.1).

CHAPTER 7. ASSESSING COMPETING SEMI-EMPIRICAL EQUATIONS

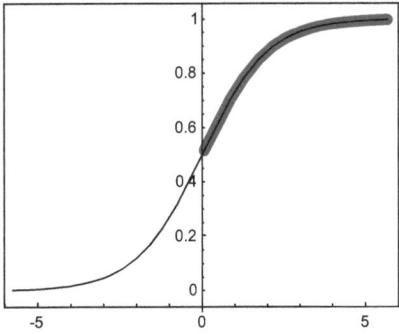

Figure 7.6: Graph of the logistic sigmoid function.

7.4.6 Equation 6: Logistic sigmoid

The characteristics of the light response are also resembled by the logistic sigmoid function. The upper right half (marked red) of the sigmoid exhibits the required behavior, see Figure 7.6: a linear initial slope, a maximum curvature in the transition phase, and leveling off to full saturation for higher values. The parameterization to model the ecosystem response to light is given by:

$$NEP(PPFD) = 2 \cdot GPP_{opt} \cdot \left(-0.5 + \frac{1}{1 + e^{\frac{-2\alpha \cdot PPFD}{GPP_{opt}}}} \right) - ER_{dayt}. \quad (7.17)$$

A literature search showed that a mathematically equivalent version, expressed as a hyperbolic tangent, has already been used for photosynthesis (Jassby & Platt, 1976):

$$NEP(PPFD) = GPP_{opt} \cdot Tanh\left(\frac{\alpha \cdot PPFD}{GPP_{opt}} \right) - ER_{dayt}. \quad (7.18)$$

Both equations have not been used with eddy covariance carbon flux data before. They are referred to as the logistic sigmoid function in the following, since this name describes its functional shape rather than its trigonometric properties.

Evaluation: This sigmoid function also exhibits the characteristics of the light response in all three phases (Figure 7.4) and the estimated physiological parameters agreed well with the ANN model results (Table 7.1). In contrast to the other light response curves, this function tends to slightly underestimate the parameters.

7.4.7 Equation 7: Exponential saturation

The cumulative function of an exponential distribution can also be used to model the light response:

$$NEP(PPFD) = GPP_{opt} \cdot \left(1 - e^{\frac{-\alpha \cdot PPFD}{GPP_{opt}}}\right) - ER_{dayt}. \tag{7.19}$$

This function has numerous names and references in the literature. The term exponential saturation function is used here, again referring to the shape of the equation.

Evaluation: Towards the limit of zero light, the slope continuously increases peaking to its highest value right at the intercept (Figure 7.4). This led to overestimates in α of 27% and ER_{dayt} of 20% (Table 7.1). This function also has no turning point. Only the third phase is well represented by this exponential function and GPP_{opt} matched the ANN model estimate within 5%.

Overall assessment

The fit performances of the seven light response curves assessed were very similar, giving little indication on the best choice. However, their mathematical properties differ significantly leading to big discrepancies in their estimates of the physiological parameters. The rectangular hyperbola, also called Michaelis-Menten equation, yielded the highest overestimates of all three physiological parameters. Though this has already been recognized (e.g. Aubinet et al., 2001; Gilmanov et al., 2003), it continues to be the most widely used light response curve in NLR algorithms, probably because of its robustness. This use should be reconsidered, since the rectangular hyperbola belongs to the wrong function class: its basic characteristics are inappropriate to describe the three phases of the light response.

The non-rectangular hyperbola exhibits the correct functional behavior only for high values of its curvature parameter. If used on data with high variability, such as the eddy carbon flux measurements over an ecosystem, the extra freedom in the curvature is not well enough constrained and leads to increased uncertainty in the estimates of the physiological parameters. Moreover, the

CHAPTER 7. ASSESSING COMPETING SEMI-EMPIRICAL EQUATIONS

regression of the non-rectangular hyperbola only converges if there is a clear curvature present in the data, which is not always the case in conditions such as dim light during the wintertime. Non-linear regression runs with the curvature parameter d fixed to an appropriate value may help to overcome this problem.

A correct description of all three phases is provided by the two sigmoid semi-empirical functions: the Smith and the logistic sigmoid. Their use ensures an appropriate basic functional form, resulting in adequate estimates of the physiological parameters even on the noisy eddy covariance data. To test their robustness, the two sigmoid functions were used with sparse data from single days, and the logistic sigmoid proved as robust as the rectangular hyperbola. Though these two equations have not been used with the eddy covariance data, a recent study in marine research by Ritchie (2008) also suggests that the logistic sigmoid function is an appropriate choice for modeling asymptotically saturating photosynthesis.

The impact of using an inappropriate semi-empirical model depends on the application. For gap-filling, the overall fit performance is crucial and all models but the linear with upper limit will be sufficient. If the model is also used to partition the data into gross primary production and ecosystem respiration or to derive the physiological properties of the ecosystem, then the choice of an inappropriate light response model can result in significant errors. To illustrate this, the half-hourly NEP measurements of Hainich for the year 2001 were gap-filled and partitioned using the NLR algorithm of Desai et al. (2005), once using the rectangular hyperbola and once using the logistic sigmoid function. The sums were calculated for the gap-filled daytime data of the productive season from May to October. The sum of NEP was almost the same, 1540 μmol CO_2 m^{-2} s^{-1} and 1550 μmol CO_2 m^{-2} s^{-1} respectively. However, the derived estimate of ER_{dayt} was 760 μmol CO_2 m^{-2} s^{-1} with the rectangular hyperbola and only 530 μmol CO_2 m^{-2} s^{-1} with the sigmoid, yielding a significant difference of 30%.

These results demonstrate the usefulness of the methodology to assess multiple hypotheses. With the aid of the data-derived light response curve, the shortcomings in some of the commonly used semi-empirical light response curves could be clearly depicted. A more detailed analysis with an extension to other types of forests is explored in Moffat (In preparation). As illustrated in this application example, the methodology can be used to revise current hypotheses by providing a link between the observations and their semi-empirical representa-

CHAPTER 7. ASSESSING COMPETING SEMI-EMPIRICAL EQUATIONS

tion. This will hopefully lead to an improved implementation of the ecosystem responses in models. In complex models, such as terrestrial biosphere models, the modeled ecosystem response is the product of many implemented processes. How the methodology can be used to characterize this synthetic ecosystem data is presented in the next chapter.

Chapter 8
Evaluating ecosystem models

The methodology provides an inverse characterization of the ecosystem response to the climatic drivers, compare also Figure 1.8. In the last chapters, this has only been applied to observed data. In the following area of application, the inverse characterization is used on synthetic data produced by two complex ecosystem models. Since the simulated ecosystem response of such complex models is the result of a combination of and interaction amongst several implemented processes, their individual dependencies are usually hard to trace. Here a direct feedback on the implementation of the processes can be provided by extracting individual aspects of the simulated ecosystem response using the presented ANN framework. Comparing the *simulated* functional relationships extracted from the synthetic data with the *observed* functional relationships extracted from the measurements, permits a comprehensive evaluation of the ecosystem models. Again the application example is demonstrated on the daytime *NEP* response at Hainich forest.

8.1 Specifying the response query: *Daytime NEP modeled by two TBMs*

The synthetic ecosystem datasets are offline runs at site level from the two terrestrial biosphere models ORCHIDEE and BETHY, further described below. The driving meteorology was based on the standardized half-hourly datasets of the Hainich forest from the Carboeurope IP database (Papale *et al.*, 2006) for the years 2000 and 2001.

Again, the study focuses on the daytime response during the active period.

CHAPTER 8. EVALUATING ECOSYSTEM MODELS

The data selection for this analysis was performed according to the description in Section 5.1, but since the dataset comprised only two (rather than three) years, the total number of half-hourly data tuples analyzed was smaller (2467).

8.1.1 Model 1: ORCHIDEE

ORCHIDEE (Organizing Carbon and Hydrology In Dynamic Ecosystems Environment) is a latest generation dynamic general vegetation model. It couples biogeophysical processes, biogeochemical processes and vegetation dynamics (Krinner et al., 2005) and has been included in global circulation model runs (Marti et al., 2005). ORCHIDEE models the half-hourly fluxes of energy, water and CO_2 with a daily update of vegetation parameters, such as leaf area index (LAI), carbon allocation and soil carbon dynamics.

The ORCHIDEE runs were performed with the standard parameterization for model evaluation purposes, without further tuning on site data. In addition to the meteorological data provided in the datasets, ORCHIDEE used soil type and vegetation composition as model inputs.

8.1.2 Model 2: BETHY

BETHY (Biosphere Energy-Transfer HYdrology) is a terrestrial biosphere model that simulates the fast vegetation processes (Knorr & Kattge, 2005). It is part of the dynamic general vegetation model JSBACH (Joint Simulation of Biosphere Atmosphere Coupling in Hamburg, Raddatz et al., 2007) and will be included in the latest version of the COSMOS earth system model. BETHY describes the short timescale processes of energy and water exchange between the atmosphere and the biosphere, including soil moisture, photosynthesis, and heterotrophic and autotrophic respiration.

The BETHY runs were performed for the gap-filling comparison project (Moffat et al., 2007). Model parameters were optimized against observed carbon and latent energy fluxes, considering prior information about parameter values to constrain these within reasonable ranges. The optimized parameter sets were then used to model the NEP response at Hainich for the whole year. As additional model inputs, BETHY used daily LAI derived from remote sensing data, soil type, texture, and depth, canopy height, and tower height.

Of the two models, BETHY has the less complex model structure. For

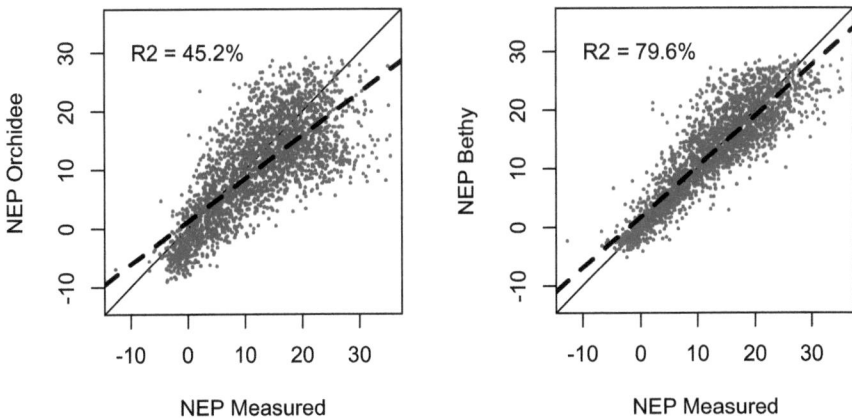

Figure 8.1: Scatterplot of the simulated *NEP* fluxes versus measured *NEP* fluxes for ORCHIDEE (left) and BETHY (right). (The thick solid line represents the linear regression fit, while the thin solid line is the identity line.)

example, the leaf area index was generated within the ORCHIDEE model itself, whereas BETHY made use of remote sensing data.

Model fit performances

The standard measure of the model fit performance is the correlation coefficient R^2. The R^2 performance of 45.2% achieved by ORCHIDEE for the selected dataset was poor, see Figure 8.1, while BETHY's R^2 performance of 79.6% was significantly better. This can to a large extent be attributed to the prior optimization of the BETHY parameters against the observed fluxes.

The fit performance only gives a number describing the mismatch between the measurements and the simulations. The ANN modeling framework is used in the following to get further insight into the characteristics of the mismatch.

8.2 Generating driver candidates

For this type of analysis, the ANN driver candidates should comprise the same set of driving inputs as used to produce the synthetic ecosystem datasets. The

CHAPTER 8. EVALUATING ECOSYSTEM MODELS

meteorological drivers of ORCHIDEE were Rg, Ta, Ts_2, Rh, SWC, and *Precip* and of BETHY were *PPFD*, Ta, VPD, and SWC. However, *Precip* was excluded from the set of driver candidates because of the lack of correlation on the half-hourly time scale, as discussed in Section 5.4. Its effect was nonetheless indirectly included by considering SWC.

Additionally, the three drivers, $PPFD_{dif}$, $PPFD_{dir}$, and f_{dif}, were added as driver candidates, since the proportion of diffuse radiation proved to be the most relevant secondary control to explain the daytime *NEP* measurements at Hainich (see Section 5.4).

The full set of ANN driver candidates was as follows:

PPFD	(Total) Photosynthetic Photon Flux Density (μmol photon m^{-2} s
Rg	(Total) Global Radiation (W m^{-2})
VPD	Vapor Pressure Deficit (hPa)
Rh	Relative Humidity (%)
SWC	Soil Water Content (%)
Ts_2	Soil Temperature at 30 cm (°C)
Ta	Air Temperature (°C)
$PPFD_{dif}$	Diffuse *PPFD* (μmol photon m^{-2} s^{-1})
$PPFD_{dir}$	Direct *PPFD* (μmol photon m^{-2} s^{-1})
f_{dif}	Diffuse Fraction (0% - 100%)

In this application example, the ANN framework is used to reconstruct the eddy flux measurements ("DATA") as well as the two TBM simulations of ORCHIDEE and BETHY.

8.3 Benchmarking with all drivers

Benchmarking with all drivers is again the first step of the characterization in Figure 8.2. For DATA, the R^2 performance was 92.2($\pm$0.1)% and the unexplained variability can be attributed mainly to noise in the measurements (see Section 5.3). For ORCHIDEE, the ANN models were able to explain 82.5($\pm$0.3)% of the variability. Since the TBM simulations are purely deterministic, the unexplained variability can be attributed to the non-instantaneous processes (e.g. lag effects) implemented in the TBM. For BETHY, the ANN models had a much higher explainability of the TBM simulations of 93.2($\pm$0.2)%.

As Abramowitz (2005) suggested, ANNs can be used quantitatively as a benchmark for the fit performance of TBMs. The ANN model could explain 92.2% of the variability in the measurements, whereas BETHY reached 79.6% and ORCHIDEE only 45.2% (see Section 8.1 above). This means that even for the tuned BETHY simulations, there is potentially over 10% more variability explainable from the climatic input drivers.

The shape of the *NEP* response in Figure 8.2 provides qualitative insight in the differences between the simulations. The measurements (black dots, left) were much better resembled by the ANN (red dots, left) and BETHY (black dots, right) than by ORCHIDEE (black dots, middle).

CHAPTER 8. EVALUATING ECOSYSTEM MODELS

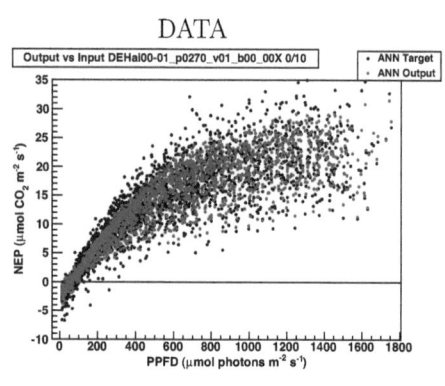

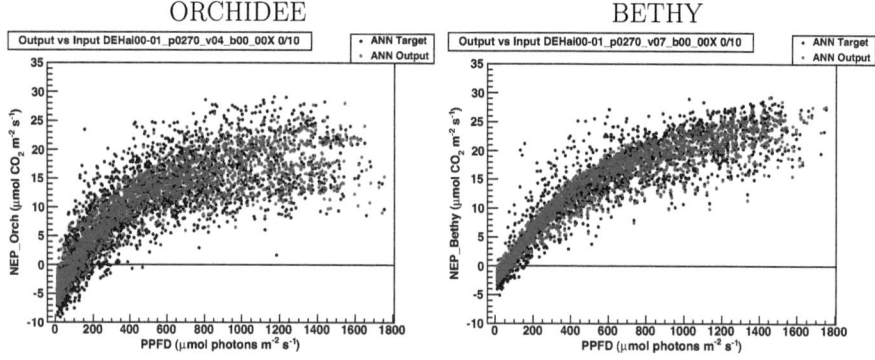

Figure 8.2: Benchmarking with all ten input drivers for the measurements (DATA), the simulations of ORCHIDEE, and the simulations of BETHY. The ANN model output (red circles) and target values (black circles) of the daytime *NEP* response are plotted against *PPFD*.

8.4 Identifying the driver hierarchy

To identify the driver hierarchy, the ANNs were trained with one climatic variable at a time on each of the datasets (DATA, ORCHIDEE, and BETHY). As expected, the radiative drivers had the highest mapping performance as primary drivers of the daytime NEP response, see Figure 8.3 (top). However, the patterns of relevance were quite different. For example, the relative ratio of the importances of radiative drivers to meteorological drivers was higher for DATA than for the simulations of BETHY and especially than for ORCHIDEE.

The difference in patterns was even more pronounced in the secondary driver analysis Figure 8.3 (bottom). The proportion of diffuse radiation (provided as $PPFD_{dir}$, $PPFD_{dif}$, or f_{dif}) was the most important secondary driver of the daytime NEP response in DATA (compare Section 5.3). Since both TBMs did not consider diffuse radiation, this variable had no explanatory power in their simulations.

A surprising trait of the ORCHIDEE simulations was the pronounced relevance of Rg as a secondary driver in addition to $PPFD$ (and vice versa, not shown). The two radiative drivers are closely related, since $PPFD$ is the photosynthetically active part of Rg (see also Figure 5.4). Therefore, Rg should add only little (if any) new information to the daytime NEP response.[1]

Looking at the meteorological drivers, Rh showed low relevance while Ta and VPD (which is a function of Ta and Rh) showed high relevances as secondary drivers in BETHY. This means that the model reacted more strongly to Ta than Rh. The high relevance of SWC in ORCHIDEE, compared to DATA, indicates a stronger impact of rain events in the model than present at the Hainich forest for these two non-drought summers. Moreover, Ta had only little relevance as a secondary driver in ORCHIDEE, a lot less than in DATA. This fact can also be seen in the next phase on the functional relationships below.

---

[1] To follow up on the discussion of light response curves in Chapter 7: The light response in ORCHIDEE is based on a non-rectangular hyperbola with a fixed curvature parameter d of 0.7, while in BETHY it is based on the Smith sigmoid function.

CHAPTER 8. EVALUATING ECOSYSTEM MODELS

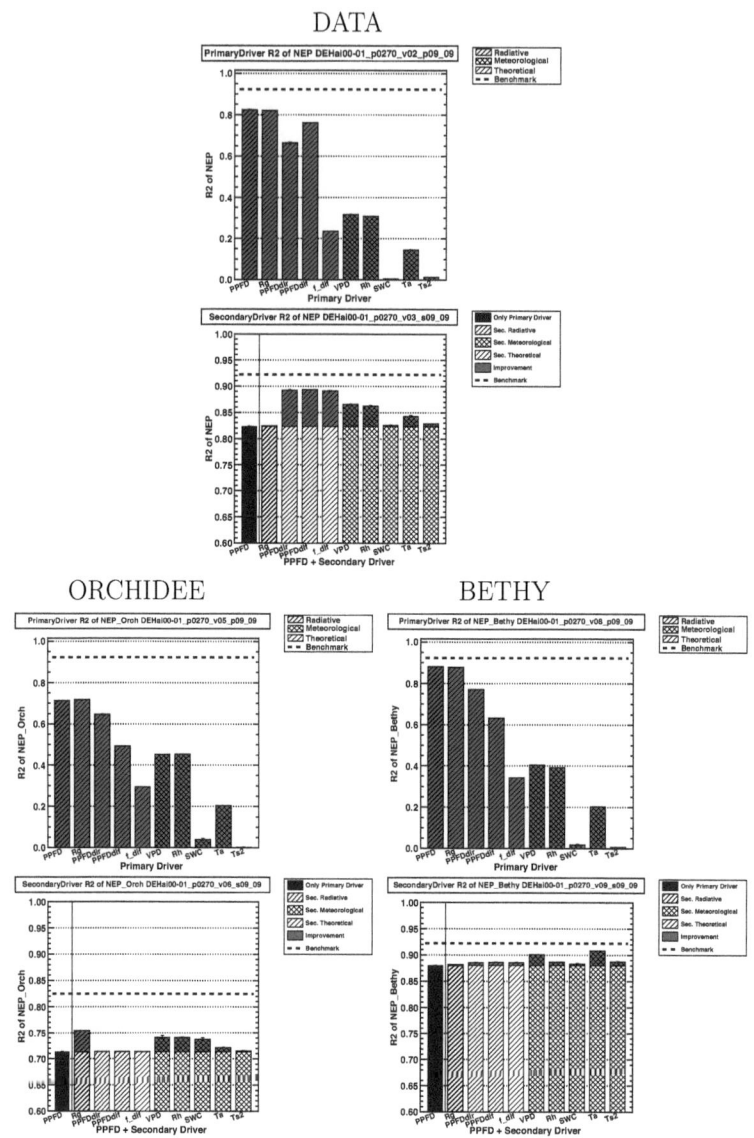

Figure 8.3: *Top:* Primary R^2 performance of the ANN models trained with a single climatic driver at a time. *Bottom:* R^2 performance of the ANN models trained with *PPFD* plus a secondary climatic driver. The performance improvement (red) indicates the relevance as a secondary driver. (See Figure 5.3 for further explanations on the graphs.)

8.5 Extracting the functional relationships

Again, the ANN framework is used to reconstruct the measurements as well as the two TBM simulations. The total radiation (provided as $PPFD$ and Rg) was the dominating climatic control for all three datasets. The most relevant meteorological drivers of DATA were the Rh, Ta, and $VPD(Ta, Rh)$, respectively, see Figure 8.3. In the following, the functional relationships of the daytime NEP response to $PPFD$ plus first VPD and then Ta are further investigated.

ANN modeling of $NEP(PPFD, VPD)$ in Figure 8.4 looked quite similar for DATA (left) and BETHY (right), whereas ORCHIDEE (middle) had a very different appearance. The VPD response is not well captured in the ORCHIDEE simulations of the Hainich forest, particularly evident in the behavior of the numerical partial derivatives (Figure 8.4, bottom middle). The ORCHIDEE relationships looked not quite as distorted when using Rg rather than $PPFD$ as the radiative input driver (not shown).

The results for $NEP(PPFD, Ta)$ behaved likewise, see Figure 8.5. DATA and BETHY were similar, both showing a negative response for high air temperature, whereas the ORCHIDEE response was flat showing no dependency on Ta.

CHAPTER 8. EVALUATING ECOSYSTEM MODELS

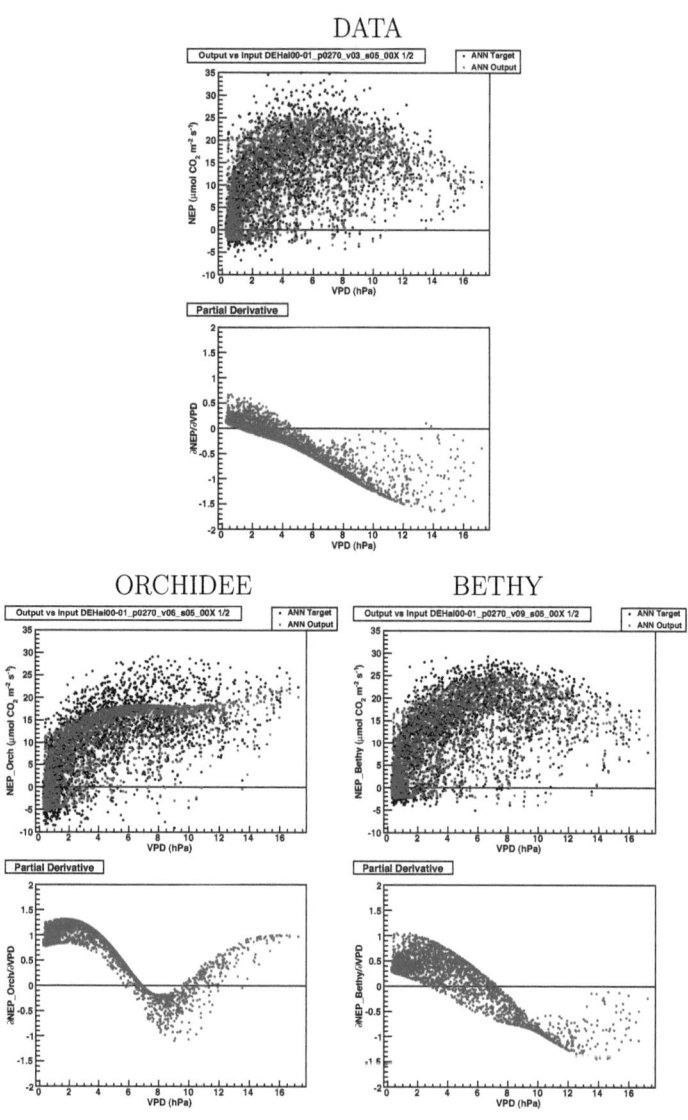

Figure 8.4: The daytime *NEP* response (top) and the numerical partial derivatives (bottom) modeled with the two climatic drivers *PPFD* and *VPD*. The response is plotted against *VPD* for the observed DATA (left), ORCHIDEE (middle), and BETHY (right).

CHAPTER 8. EVALUATING ECOSYSTEM MODELS

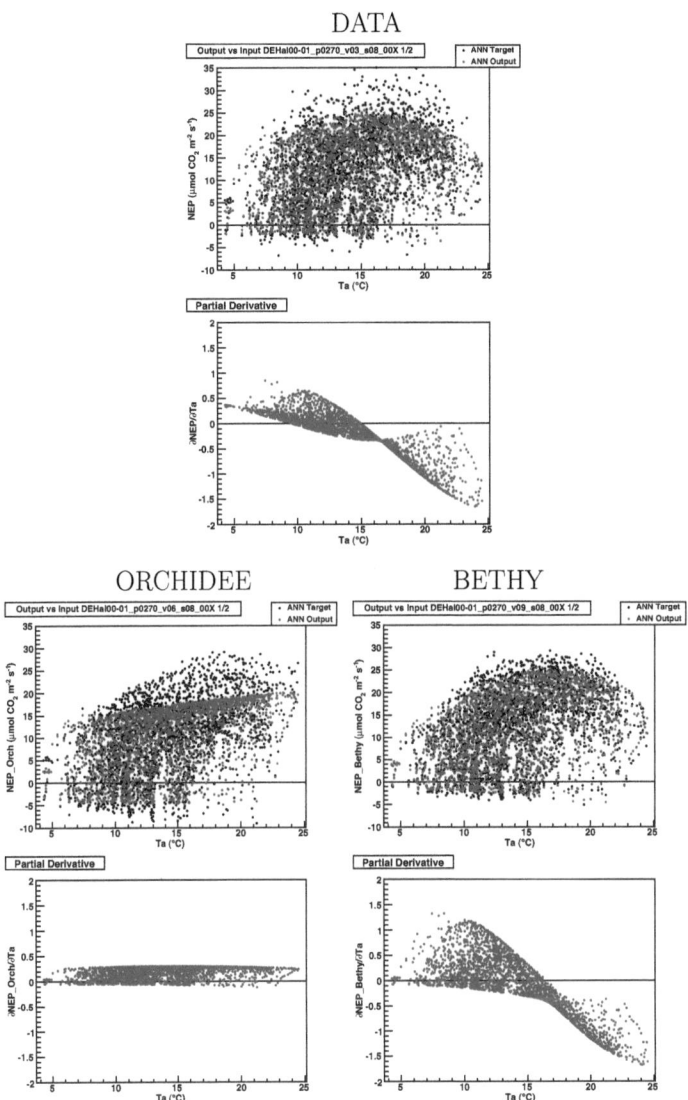

Figure 8.5: The daytime *NEP* response (top) and the numerical partial derivatives (bottom) modeled with the two climatic drivers *PPFD* and *Ta*. The response is plotted versus *Ta* for the observed DATA (left), ORCHIDEE (middle), and BETHY (right).

8.6 Discussing the induced hypotheses

The characterization of the hierarchy of the climatic drivers as well as the functional relationships of ORCHIDEE and BETHY discloses *how* the daytime *NEP* response is simulated by the TBMs. By comparing these to the relationships extracted from the measurements (the "truth"), the two TBMs can be evaluated.

Overall, BETHY showed a good representation of the daytime NEP response at Hainich, but the implementation of diffuse radiation is highly recommended to capture the full response at this site. This might improve the model to explain the extra 10% of the variability in the measurements that can be captured by the benchmark ANNs on DATA.

The poor R^2 performance of simulated versus measured data already indicated some deficits in the ORCHIDEE simulations. The hierarchy of the climatic controls depicted an inexplicable importance of *Rg* and *PPFD* as each other's secondary drivers. The unusual dependency on *VPD* and the missing dependency on *Ta* of ORCHIDEE are an indication that internal model dynamics must play a key role in causing distortion and cancellation of the effects. This is in agreement with results of Abramowitz *et al.* (2007), who found that the bias errors of three different TBMs, including ORCHIDEE, were highly systematic (by using self-organizing maps modeling the mismatch between simulation and observations).

Hence, the ORCHIDEE simulations do not reproduce some of the critical aspects of the daytime *NEP* at the Hainich forest. It would be interesting to see how tuning of the run would alter the results. A newer version of ORCHIDEE (Zaehle & Friend, 2010) can be run with the diffuse radiation. This might also help to better resemble the relationships found in the observed DATA.

CHAPTER 8. EVALUATING ECOSYSTEM MODELS

This application example showed that the presented methodology can be used not only on observed but also on synthetic ecosystem datasets. The benchmark performance, the derived hierarchy of the climatic drivers, and the extracted functional relationships provided a wide variety of criteria for the evaluation of two ecosystem models. This area of application also offers the perspective to perform model-data synthesis (e.g. Raupach *et al.*, 2005; Wang *et al.*, 2009) by systematically varying the ecosystem model structure in order to gain information about the internal model dynamics. How the synthetic ecosystem datasets produced by the ANN models themselves (the ANN model output) can be used for interpolating missing data in the measurements, is subject of the next chapter.

CHAPTER 8. EVALUATING ECOSYSTEM MODELS

Chapter 9
Interpolating missing data

One of the drawbacks of the eddy flux measurements is their fragmentation (see Section 1.2.3). To calculate daily and annual sums, these measurement gaps need to be filled. The sums are of widespread interest, e.g. for estimations of the ecosystem carbon budgets, for the evaluation of TBM simulations, or for scaling to biometric measurements. Several gap-filling techniques, including ANNs, have been developed to interpolate the missing *NEP* data. In a comprehensive comparison of fifteen gap-filling techniques (Moffat *et al.*, 2007), the performance of the techniques was evaluated by comparing observed *NEP* with predicted (filled) *NEP* values on artificial gaps.

As an additional application, the methodology can also be employed as a gap-filling technique. The relationships of the ecosystem response to all the climatic drivers mapped in the benchmark ANN models permit the reconstruction of missing *NEP*, as discussed in the following chapter. To compare the performance to the other fifteen gap filling techniques of Moffat *et al.* (2007), the same "keyfile" for the artificial gaps and one of the "golden files", the Hainich dataset from the year 2000, were used. In contrast to the previous application examples, not only the summer daytime data but the full yearly datasets were analyzed.

9.1 Specifying the response query: *Gap-filling of missing NEP*

The golden file contains the *NEP* measurements and an appointed set of associated meteorological data. Gaps in the meteorological data were previously filled

(Moffat et al., 2007). Only the best quality *NEP* measurements were provided, resulting in ~20% gaps in the daytime data and ~65% gaps during nighttime in the Hainich 2000 dataset.

Secondary datasets with artificial gaps scenarios are generated by flagging 10% of the data as unavailable, i.e. artificial gaps. Ten percent was chosen as a compromise between sufficient power for statistical analyses and avoiding excessive additional fragmentation of the data files. To achieve statistical validity, the artificial gaps were distributed randomly and each artificial gap length scenario was permuted ten times, thereby sampling $100\% - (100\% - 10\%)^{10} = 65\%$ of the total yearly data. The artificial gaps are superimposed on the already incomplete data in the files, without regard for the distribution of real gaps in the *NEP* data.

Different types of artificial gaps from "very short" with a gap length of single half-hours to "long" with a gap length of ten days were considered in the gap filling comparison (Moffat et al., 2007). To condense the following analysis, only the ten very short artificial gap scenarios plus the real gaps in the observed *NEP* data were filled.

Daytime and nighttime differentiation

Daytime refers to positive photosynthetic photon flux density ($PPFD > 0$) and nighttime refers to periods of the day with no light ($PPFD = 0$). For determining the gap-filling performance, the weighting of the daytime and nighttime contributions to the statistical metrics is incorrect when day and night are taken together. More precisely, the ratio of the number of daytime to nighttime gaps for the real gaps is at odds with the day-night ratio of the artificial gaps. The percentage of available observed *NEP* is 80% for daytime and 35% for nighttime data. Thus, the distribution of real gaps of 20% daytime to 65% nighttime results in a day-night ratio of approximately 1:3. By contrast, the secondary datasets have ten percent artificial gaps resulting in 8% daytime and 3.5% nighttime gaps, a ratio of approximately 2:1. Therefore, the analysis of the gap filling results was performed separately for daytime and nighttime periods.

For the ANN training, the data was also grouped into subsets of daytime and nighttime data. The separately trained ANN models performed better than the ANN models trained on the full yearly dataset, even when a step function was included as an additional input driver to allow the ANNs to distinguish between day and night. This might be due to the imbalanced ratio of daytime to nighttime training data, since the full yearly dataset contains more than double the amount of daytime tuples. Besides, the absence of photosynthesis changes the underlying biological processes caused by the same input drivers.

9.2 Generating driver candidates

The meteorological variables provided in the golden file are *Rg*, *PPFD*, *Ta*, *Ts*, *SWC*, *Rh*, and *Precip*. Since only *PPFD* or *Rg* are needed, the ecophysiologically more meaningful *PPFD* was chosen. *Precip* has no direct effect on the half-hourly time scale (see Section 5.4) and was therefore excluded. Two derived latent variables, *OptProd* and *dOptProd*, were added to provide information about the state of the ecosystem and are further explained below.

The input driver set used thus included the following variables:

PPFD	Total *PPFD* (μmol photon m^{-2} s^{-1})
Ta	Air Temperature (°C)
Ts	Soil Temperature (at 30 cm) (°C)
SWC	Soil Water Content (%)
Rh	Relative Humidity (%)
OptProd	Maximum daily value of *NEP* smoothed over $\sim$1 month
dOptProd	Derivative of *OptProd* smoothed over $\sim$3 months

Variables providing information about the state of the ecosystem are necessary if the gap-filling is performed on yearly data. Their effects were demonstrated by training the ANNs with *PPFD* plus one of the potential ecosystem state variables, see Figure 9.1. The panels each show the course of the year: the left two for the input drivers and the right panels for the *NEP* measurements (black) and the ANN model output (red).

When only *PPFD* was used as an input driver, the shape of the ANN model output was flat (top row). Adding any of the ecosystem state drivers changed the shape of the ANN model outputs to more closely resemble the yearly *NEP* response, which is mainly driven by the phenology of the forest. The slowly varying soil temperature *Ts* improves the results (second row), though the exact features, e.g. the dropping of *Ts* in the summer, have no direct bearing on the ecosystem *NEP* response.

Bell-shaped fuzzy variables can be used to describe the seasons (spring, summer, fall, winter). Although the mapping of the ANN model with only one

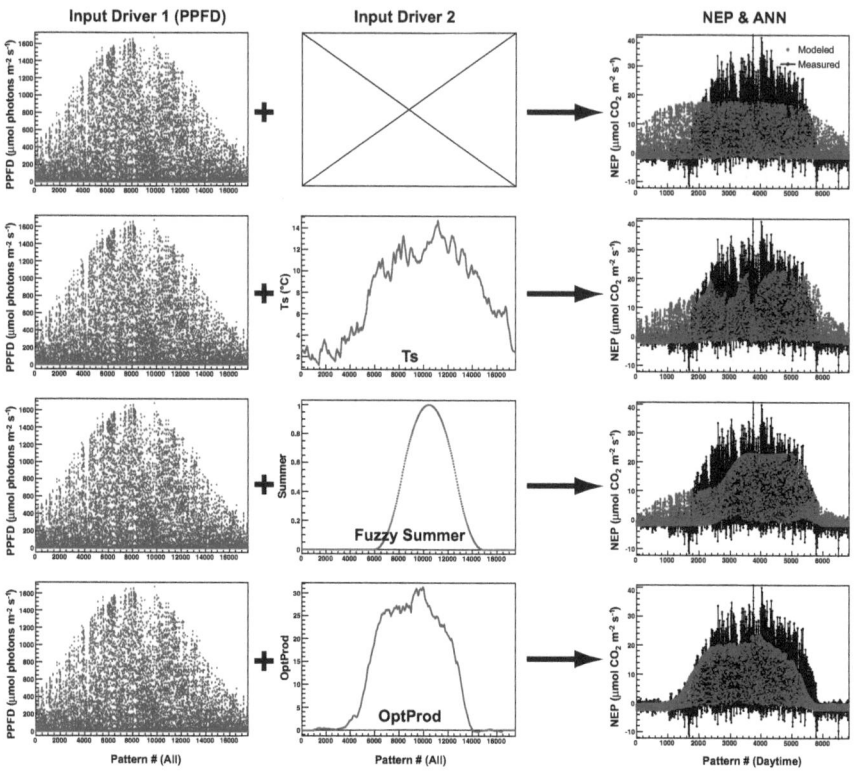

Figure 9.1: Effect of input variables describing the yearly ecosystem state on the modeled *NEP* output. (In the two left columns, all 17520 yearly half-hours are plotted, whereas in the right column only the subset of the daytime data is shown.)

of these fuzzies, "Fuzzy Summer" (third row) led to a delayed timing, the full set of four fuzzies offered enough freedom to capture the main shape of the response (not shown).

Almost the same performance as with all four season fuzzies was obtained with only one variable, the latent variable *OptProd* derived from the fragmented *NEP* data. *OptProd* (bottom row) describes the maximum daily value of *NEP* smoothed by a moving average over ~1 month and can be regarded as a measure of the optimum productivity. The ANN model using *OptProd* was able to capture the basic shape of the response. For the ANN training on nighttime data, the derivative $dOptProd$ was of importance, since it provided information about whether the ecosystem is in its spring onset or autumn senescence phase.

Preferably, a direct measure of the phenology, such as LAI derived from satellite data, should be used if available. ANN models using LAI data directly were best at reproducing the yearly course (not shown).

Unfortunately, the golden file does not contain a variable describing the proportion of diffuse radiation. Since it proved to be the second most relevant driver to explain the summer daytime NEP in Section 5.4, $PPFD_{dif}$ was used in an extra gap-filling run to determine its effect on the gap-filling performance.

9.3 Benchmarking with all drivers

The performance of the ANN models trained with all climatic drivers is shown in Figure 9.2. The explainability of the variability in the dataset was much higher during daytime with an average R^2 of 87.5%. At nighttime, the signal to noise ratio is much lower (see also Moffat et al., 2007) and the R^2 was only 52.7%.

The bias error mainly depended on the placement of the artificial gaps in the dataset. This can be deduced from the standard deviation between ANN training permutations being much smaller than the absolute magnitude of the bias.

Due to the smaller NEP flux magnitudes at nighttime, $SDev$ was lower than during daytime. The remaining error can again be attributed to the random error (noise) in the measurements (Moffat et al., 2007; Richardson et al., 2008, and see also Section 5.3). The effect of filtering the noise by gap-filling can be seen in Figure 9.3, showing the results for modeling the whole time series with only the real gaps. The ANN model output (red line) is much smoother than the original measurements (black line). The modeled ANN output can now be used to fill the real gaps in the measurements.

9.4 Discussing the induced hypotheses

Figure 9.4 shows the $RMSE$ performance of the ANN framework presented here (red symbols) in comparison to the other fifteen gap filling techniques of Moffat

CHAPTER 9. INTERPOLATING MISSING DATA

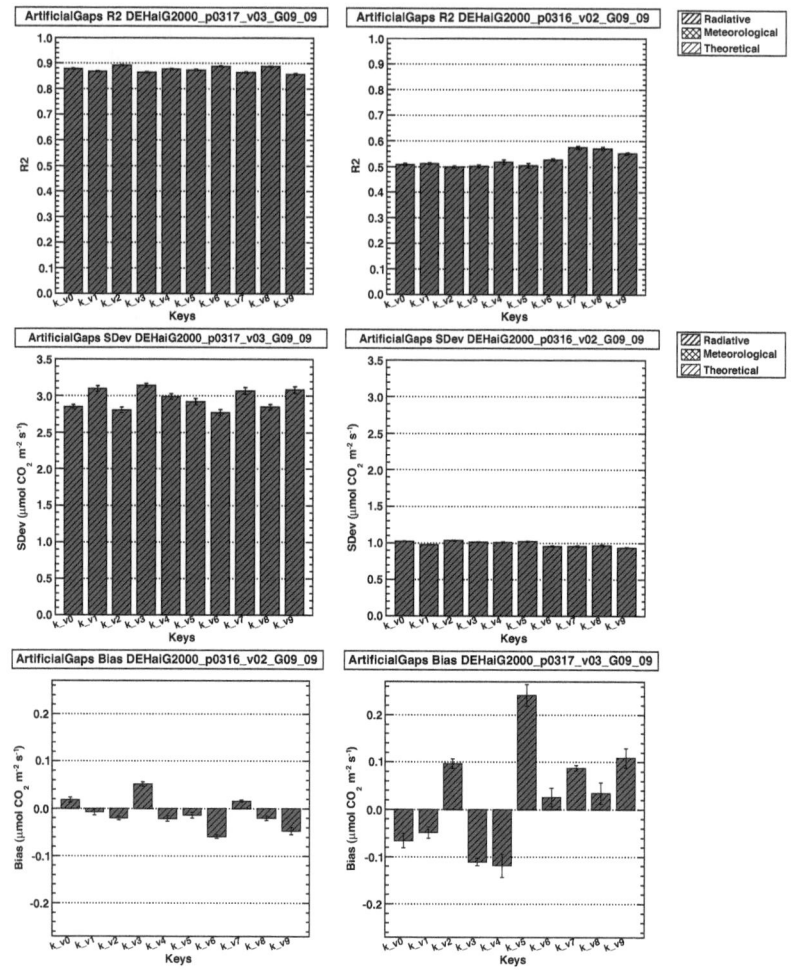

Figure 9.2: R^2 (top), *SDev* (middle), and *Bias* (bottom) performance for each of the ten very short artificial gap scenarios. The benchmark ANN models were trained separately for daytime (left) and nighttime (right). (The error bars show the standard deviation of ten ANN training permutations.)

et al. (2007). The *RMSE* was as low as of the other three ANN techniques examined. Including $PPFD_{dif}$ reduced the daytime *RMSE* by almost 20% and

CHAPTER 9. INTERPOLATING MISSING DATA

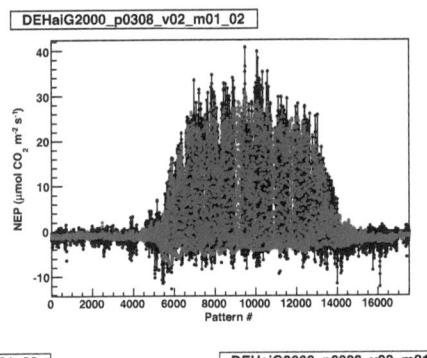

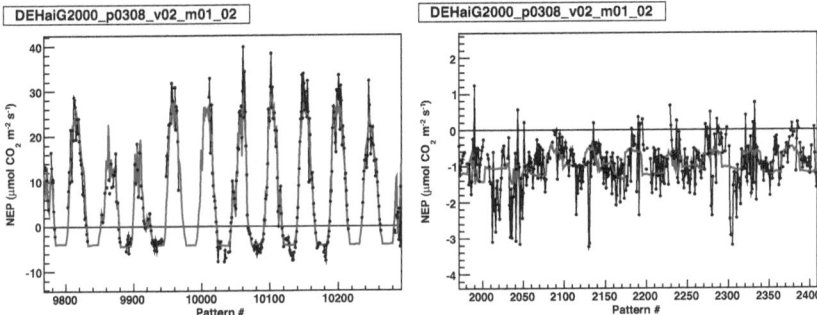

Figure 9.3: Half-hourly *NEP* measured (black) and modeled (red), plotted for the entire year of Hainich 2000 (top), ten summer days (bottom left), and ten winter days (bottom right).

led to the best performance of all. This stresses again the importance of the diffuse radiation as a driver of the daytime *NEP* response. Hence, it is highly recommended to account for $PPFD_{dif}$ in gap-filling routines.

For the prediction of the annual sum, a low systematic bias error is an important criterion for a gap-filling technique (Moffat et al., 2007). The deviation of *Bias* between ANN training permutations was small, much smaller than its absolute magnitude, see Figure 9.2 (right). This indicates that the implementation of the epoch smoothing (see Section 2.5) indeed fixed the problem of large biases in the earlier version of this ANN framework used in Moffat et al. (2007). There, the bias deviations had been of the order of a factor of ten larger.

This last of the five areas of application demonstrated the use of the methodology as a gap-filling technique. The phase of identifying the driver hierarchy

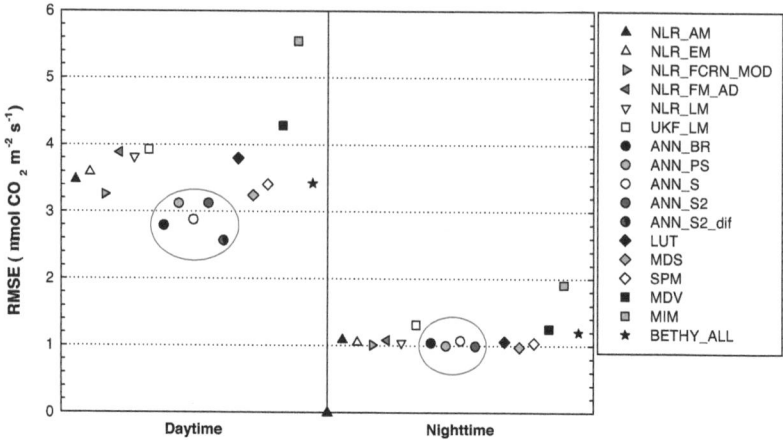

Figure 9.4: *RMSE* performance of the gap filling techniques analyzed in Moffat *et al.* (2007), averaged over the ten very short gap scenarios at Hainich 2000 and separated into daytime (left) and nighttime (right) data. Marked in red are the ANN results of the modeling framework presented here, without (ANN_S2) and with $PFFD_{dif}$ included (ANN_S2_dif).

can additionally be used to optimize the selection of input drivers for the ANNs, as in the case of diffuse radiation. The underlying ANN framework met the criteria of a "good" gap-filling technique according to the comprehensive gap filling comparison by Moffat *et al.* (2007). This also emphasizes the reliability of the modeling results, which is investigated further in the next chapter. With the gap-filling capability, the methodology again showed its strength and breadth as an instrument to analyze and use the information hidden in the large, fragmented, and noisy ecosystem datasets.

Part III
Reliability and conclusions

Chapter 10
Reliability of the modeling results

The methodology is based on an inductive modeling framework where the functional relationships are derived exclusively from the data. This chapter explores the reliability of the modeling results obtained in the application examples.

One basic requirement for the reliability of such purely empirical models is their good generalization beyond the dataset used for training. This generalization ability is tested with the cross-validation method in Section 10.1. The effect of weight regularization on the generalization is shown for one example in Section 10.2. How the uncertainties in the inputs impact the derived functional relationships is explored in a sensitivity study in Section 10.3. The chapter closes with on overall assessment of the reliability of the modeling results in Section 10.4.

10.1 Cross-validation

One of the criteria of a good function approximation is the ability of the trained ANN model to fit new (unknown) data. The opposite is an overfitted model, where the network has learnt the training set perfectly but interpolates unknown points erroneously (Rojas, 1996). Overfitting generally occurs when a model has too many degrees of freedom. Simple and plain functional relationships are usually more robust. In the ideal case, the derived functional relationships represent only the universally valid information (causal relations) and not the irrelevant information (noise) in the dataset.

CHAPTER 10. RELIABILITY OF THE MODELING RESULTS

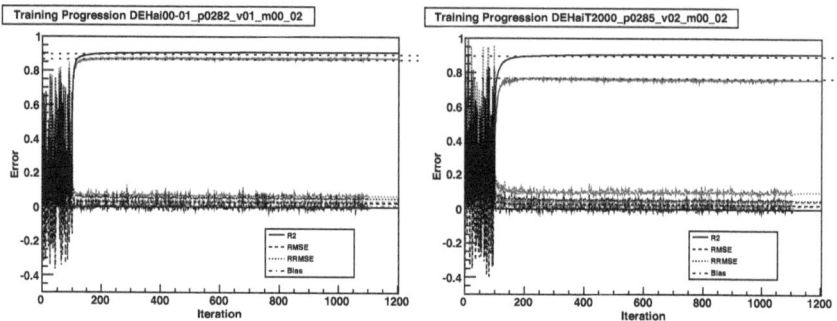

Figure 10.1: Error progression during ANN training on Hainich data from the year 2000. The training process was cross-validated with data from a different year (2001, left) and a different site (Tharandt, right). (The graphs show several error measures of the training dataset in black and of the cross-validation set in red. The thin dashed horizontal lines are to guide the eye.)

For noisy training data, the goal of a good generalization contradicts the objective of a minimization of the error function, since the ANN model should not learn the noise. Hence if the model has good generalization properties, its performance is likely to be adversely affected.

Several procedures have been implemented in the modeling framework to avoid overfitting and ensure good generalization (Chapter 2):

- Keeping the number of initial layers and hidden nodes small in order to decrease the plasticity of the network;
- Early stopping of the training as soon as the error levels off;
- Pruning of the nodes in the network;
- Penalizing large weights by regularization.

A common method to determine the generalization ability is to train the network on only part of the dataset and use the rest for cross-validation. With increasing specialization only of the training dataset, the network performance on the cross-validation set would decrease.

In the case of the carbon flux measurements, the data is highly repetitive and subsets are not independent. To get a more independent dataset, data from a different year was first used for cross-validation. The left graph in Figure 10.1

shows an example, where the daytime summer *NEP* response of the Hainich forest to the climatic drivers *PPFD*, *VPD*, *Ta*, *Ts2* was trained on the data from the year 2000 (black line), and cross-validated with data from the year 2001 (red line). There is no degradation in performance with increasing number of learning iterations visible in the cross-validation set, even though the training process was continued without pruning.

This test was repeated using a dataset from a different site, Tharandt, for cross-validation. Tharandt is an evergreen needleleaf forest about 250 km east of Hainich. Here, again, although the overall performance of the cross-validation set in the right graph of Figure 10.1 is lower, there is neither further improvement nor a significant degradation. Thus even cross-validation with a different site showed no evidence of the expected degradation with increasing training iterations, i.e. the further specialization (learning) on the Hainich dataset did not happen at the expense of the generalization to the Tharandt dataset.

A possible explanation is that the dataset is still not independent enough, though Tharandt is a different ecosystem type (evergreen needleleaf versus deciduous broadleaf forest) in a different region. However, it is more likely that the ANN training is not overfitting the dataset in the first place, since the over 1000 data tuples are trained on a network with less than ten hidden nodes (little freedom). Furthermore, the input as well as the output data is subject to inherent random noise, see properties of the eddy flux measurements in Section 1.2.3 and estimates of the uncertainty in Section 4.3. This means that even measurements of the same carbon flux under the same meteorological conditions would result in slightly different input and output values. The set of noisy data tuples presents an inconsistent learning task for the ANN with a minimum of the error function for the most general behavior, thus a network with a good generalization; see also the discussion on the impact of input uncertainty below (Section 10.3).

10.2 Example for regularization

To show the effect of adding the weight regularization term to the error function of the backpropagation algorithm (Section 2.2), the cross-validation test with Hainich and Tharandt from the last section was repeated using soil water content *SWC* as an additional input driver. *SWC* has a very different range in the two extracted dataset: 36% to 46% at Hainich and only 9% to 11% at

CHAPTER 10. RELIABILITY OF THE MODELING RESULTS

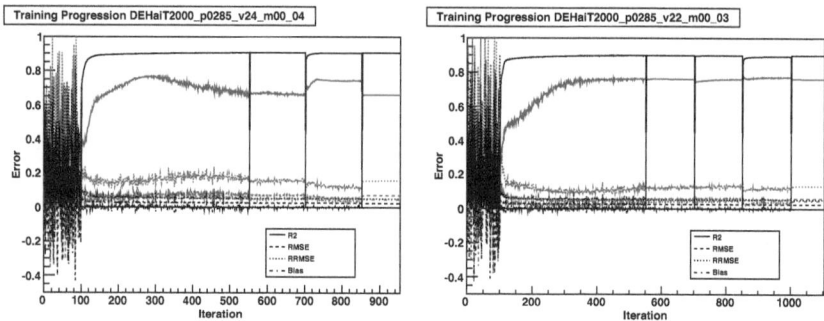

Figure 10.2: Error progression during the ANN training on Hainich data from the year 2000 using the *SWC* as an additional input driver. The training process without (left) and with (right) weight regularization was cross-validated with Tharandt data. (The graphs show several error measures of the training dataset in black and of the cross-validation set in red.)

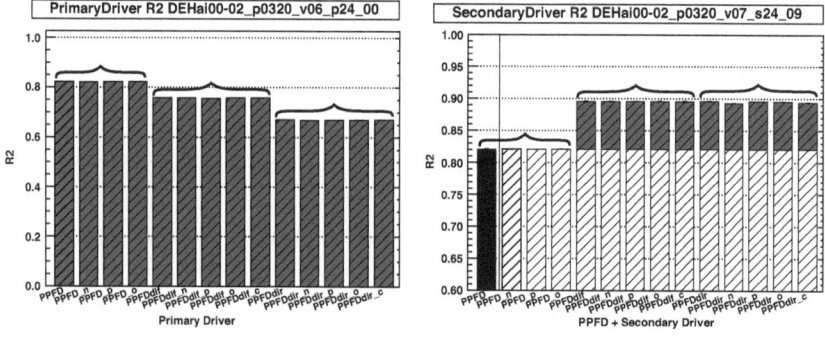

Figure 10.3: Primary (left) and secondary (right) R^2 performance of the ANN models trained with three different setups of artificial noise added to the input drivers.

Tharandt.
Since the ANN was trained on the higher Hainich *SWC* range, the extrapolation to the much lower Tharandt *SWC* values during cross-validation resulted in an irregular performance, see Figure 10.2, left. After activating the regularization term that penalizes large weights, the network performance on the cross-validation dataset was more stable and remained high, thus showed a more consistent and improved generalization, see Figure 10.2, right.

10.3 Impact of input uncertainty

The generalization of the response modeled by the ANNs depends not only on the noise in the output (target) data but also in the input data. The following are error specifications of the instrument devices used at the Hainich flux tower (Olaf Kolle, personal communication):

- *PPFD*: relative error of ±5%,
- *Rd*: relative error of ±2%,
- *Rh*: relative error of ±4%,
- *Ta*: ±0.2°C, and
- *Ts*: ±0.2°C.

To assess the impact of the input uncertainties on the network performance and on the functional relationships, a sensitivity study was performed by introducing artificial uncertainties on the radiation measurements *PPFD*, $PPFD_{dir}$, and $PPFD_{dif}$. The maximum relative error of 5% (see instrument specifications above) was chosen as the range of uncertainty. The following artificial scenarios were considered:

1. Uncorrelated random noise (n) of up to 5%,
2. Positive (p) and negative (o) offset of 5%, and
3. Correlated random noise (c) for $PPFD_{dir}$ and $PPFD_{dif}$ of up to 5%.

The study was performed with the same dataset as in Chapter 5. Figure 10.3 shows the plot of the primary and secondary R^2 performance. The ANNs with or without an artificial noise added yielded the same performance, e.g. the bars for *PPFD*, $PPFD_n$, $PPFD_p$, and $PPFD_o$ all have the same height.

To investigate the impact on the derived functional relationships, the ANN models were trained with correlated noise on $PPFD_{dir\_c}$ and $PPFD_{dif\_c}$. The modeled response and even the derivatives in Figure 10.4 (with noise added) closely resemble those in Figure 5.11 (without noise). There are always slight differences, since one ANN model is never identical to another ANN model due to the randomly initialized weights and the randomly chosen order of the data tuples during training. However, the basic shape of the relationship (form, offsets, magnitude) is not affected. The same is true for the two other artificial noise scenarios (not shown).

CHAPTER 10. RELIABILITY OF THE MODELING RESULTS

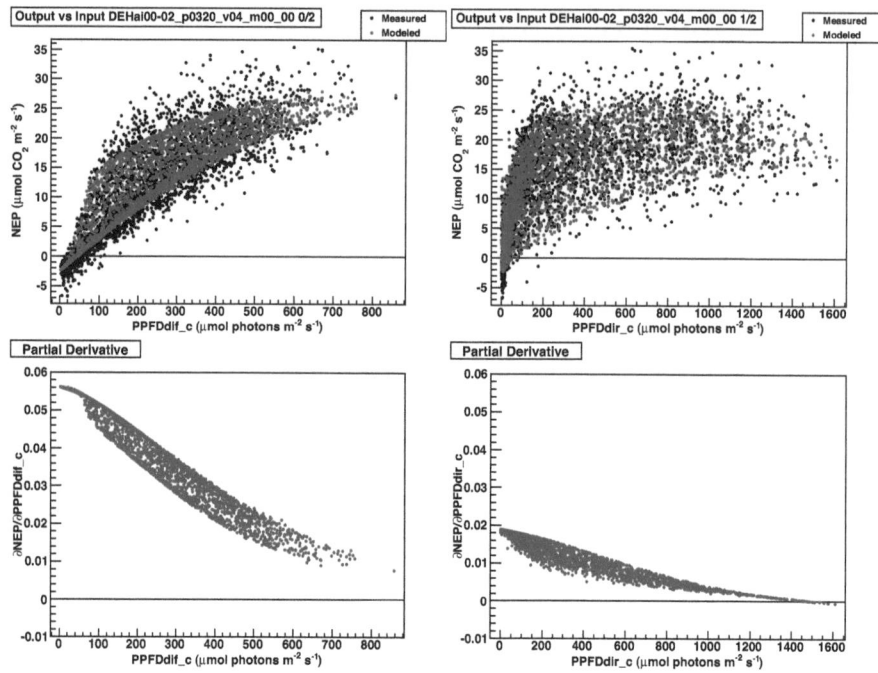

Figure 10.4: ANN model predictions (top) and numerical partial derivatives (bottom) of the daytime *NEP* response modeled with noisy $PPFD_{dif\_c}$ and $PPFD_{dir\_c}$.

None of the tested scenarios (positive or negative offset and uncorrelated or correlated random noise) had a significant impact on the network performances or on the functional relationships. Indeed, adding noise to the inputs is sometimes used as a technique to discount an overfitted response and improve the generalization (Reed & Marks, 1999). (See also discussion on noise in Section 10.1 above.)

10.4 Overall assessment

Purely empirical models with a good generalization will yield reliable results, i.e. high explainability, robustness, and plausibility. The reliability of the presented

CHAPTER 10. RELIABILITY OF THE MODELING RESULTS

modeling results is assessed for the following five criteria:

Mapping performance: The mapping performance gives a measure how well the ecosystem response can be reproduced by the model. The benchmarking results of Section 5.3 attest a high mapping performance of the ANN models. The standard deviation of the ANN model residuals was even lower than previous estimates of the random noise for the same dataset. These findings are supported by other studies (e.g. Abramowitz, 2005; Moffat et al., 2007), where ANN models have outperformed classical semi-empirical methods. Furthermore, the modeling framework fulfilled the criteria of a good gap-filling technique in Chapter 9.

Robustness of training permutations: Each training process is different due to the random initialization of the weights, the random pattern shuffling, and thereby variable node pruning. This leads to differences in the final node structures and weight parameters of the ANN models even for the same training dataset. However, the performance of the ANN models should be independent from the individual training process. To get a measure of the robustness, each training was repeated ten times. The obtained standard deviation over ten ANN training permutations is almost invisible in all of the primary, secondary, and tertiary R^2 performance plots, and is only a little larger but still small for the $SDev$ and sensitivity plots (see respective figures throughout this manuscript). The small impact of the training permutations attests to the robustness of the ANN training.

Robustness of driver relevance: Another important aspect is the sensitivity of the ANNs to the way a climatic driver is provided in the dataset. In the case of the daytime response, the second most relevant information next to the total light was the proportion of diffuse light. This information was presented to the ANN models as total plus diffuse radiation, total plus direct radiation, total radiation plus diffuse fraction, or as direct plus diffuse radiation (see Section 5.4 and 5.5.2). Each of these combinations yielded the same R^2 network performance, demonstrating the outstanding ability of the ANNs to extract the correlations of the input drivers with the responding output even when the inputs change units, magnitude, or are (non-)linearly transformed. The independence from the representation

CHAPTER 10. RELIABILITY OF THE MODELING RESULTS

of the input drivers also proves the applicability of the ANN mapping performance as a measure of the relevance of the climatic drivers.

Robustness of functional relationships: A good generalization beyond the training dataset also implies that the derived functional relationships remain (almost) the same for slightly different training data. One example is given in Section 10.3 above, where the functional relationships were robustly derived, even with artificial noise added to the inputs. Since the two ANN models (with and without noise) were trained separately, this also indicates that the derived functional relationships were independent of the individual training process.

Plausibility of functional relationships: Last but not least, the functional relationships should be not only robust but also plausible. The plausibility of the functional relationships extracted in this study are discussed throughout the manuscript. For example, the purely empirical functional relationship obtained for the light response in Chapter 7 agrees well with current plant physiological hypotheses. Though this might be expected, it is not the case for some of the other commonly used semi-empirical light response curves. Other examples are the results of the high light use efficiency of diffuse radiation in Section 5.5.2 or the sensitivity to vapor pressure deficit in Section 5.6. Again, these results are supported by the literature.

The modeling framework performed well for each of the five criteria and can hence be assumed to produce reliable modeling results. The high reliability also points to and affirms the capability of artificial neural networks to be used in an inductive manner as a glass box.

The examined aspects, the cross-validation, the regularization example, the sensitivity study of the input uncertainty, and the overall assessment, all attested to a good generalization ability and thus a high reliability of the modeling results. However, a good generalization of the relationships present in the training dataset does not necessarily mean that these relationships have universal generality. Guidance for setting up the dataset to be representative for the queried ecosystem response and for interpreting the purely empirical models to induce hypotheses is provided in Chapter 4. These guidelines need to be taken into consideration when using an inductive modeling approach to derive ecophysiological properties.

Chapter 11
Conclusions and outlook

Ecosystem datasets are usually so complex, noisy, and even fragmented that the underlying causalities cannot be obtained just by visual evaluation of the measurements. To extract the causalities *with as little prior information as possible*, an inductive methodology was developed. The methodology is based on artificial neural networks and allows an inverse mathematical characterization of the ecosystem response to the climatic controls directly from the data. The inverse characterization can be applied to both observational as well as synthetic datasets.

The different areas of application of the methodology were all demonstrated for the same ecosystem response, the daytime net carbon response of the Hainich forest:

- The comprehensive characterization identified the climatic controls of the ecosystem response as well as their underlying functional relationships and sensitivities. The often neglected diffuse radiation turned out to be the second most important driver of the daytime response.

- The hypothesis regarding the net effect of the diffuse radiation on net carbon flux was tested using two theoretical drivers. The testing supported previous conclusions obtained with a biophysical model that there is an optimum range for low fractions of diffuse light.

- Seven semi-empirical equations for modeling the light response were assessed with the sigmoid curve showing the best matching characteristics.

- Synthetic data from two complex prognostic terrestrial biosphere models was characterized in comparison to the relationships present in the

CHAPTER 11. CONCLUSIONS AND OUTLOOK

observational data, revealing some significant shortcomings in the model simulations.

- The use of the methodology as a gap-filling technique provided reliable estimates of the missing net carbon flux data.

The data-derived physiological properties corroborated existing hypotheses or indicated new or different features present in the data. The wide range of application areas and their ecophysiological relevance shows the capability of the methodology to serve as a key instrument for analyzing ecosystem datasets. The results also highlight the benefit of the methodology to provide a new link between the observations and their semi-empirical representations in the modeling world. By supplying this link, this inductive methodology is complementary to the classic hypothetic-deductive approach, thereby furthering the understanding of the underlying processes as well as promoting their implementation in models. This, in turn, will help in predicting the effects of changing environmental conditions on the terrestrial biosphere.

The ecosystem of a mature beech forest in a temperate climate, the Hainich forest, was chosen as the common basis of the application examples. The worldwide network of eddy flux towers in FLUXNET offers the opportunity to investigate ecosystems ranging from the arctic to the equatorial savannah. Applying the inductive methodology to the various land vegetation types and climate zones will advance the understanding of the net carbon fluxes between the terrestrial ecosystems and the atmosphere. For managed ecosystems, the ability to include theoretical variables, such as a fuzzy variable to describe the harvesting event, will be of benefit. The theoretical variables also offer the possibility to include time lag effects and to determine their relevance for the ecosystem response.

The methodology is not limited to the net carbon flux but can be extended not only to carbon flux partitioned into gross primary production and respiration fluxes, but also to the energy and momentum fluxes, or to other greenhouse gases. Another field is its use as a model-data fusion tool to study the behavior of more complex models such as terrestrial biosphere models. Hence, there are a wide variety of potential applications and I am looking forward to pursuing them.

Appendix A
Technical implementation

The technical implementation of the methodology required:

1. Adaptation of ANNs as an inductive modeling framework (Chapter 2),
2. Implementation of the data analysis tools (Chapter 3), and
3. A highly flexible data setup to specify the response query and generate the driver candidates with automated processing routines for the different phases (Chapter 4).

The required versatility and flexibility was achieved with object-oriented programming in C++. The program makes use of the extensive CERN-ROOT libraries (Brun & Rademakers, 1997) for file input/output of the class instances as well as for producing the 2D and 3D graphics. ROOT also comes with a C++ line interpreter called CINT, which permits flexible prototyping and direct code execution.

The following sections describe the object-oriented network implementation (Section A.1), the program structure (Section A.2), the program flow (Section A.3), and give an example of a typical pattern generation script (Section A.4) and run script (Section A.5).

A.1 Object-oriented network implementation

In contrast to many purely numerical neural network algorithms, the implemented algorithm is fully object-oriented with each artificial neuron (node) and each link representing an individual entity (after Rogers, 1996). This object-oriented approach has several advantages, such as:

APPENDIX A. TECHNICAL IMPLEMENTATION

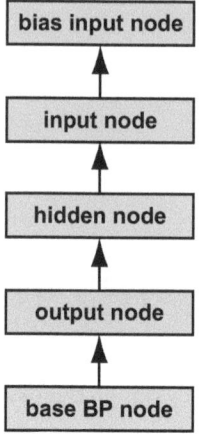

Figure A.1: Inheritence scheme of the backpropagation node classes.

- The topology of the network is fully flexible in layers, nodes, and their connections;
- Each node and each link can in principle have different attributes and behavior;
- The node and link classes can be implemented with inheritance.

The object-oriented implementation of the backpropagation algorithm (Section 2.2) directly corresponds to the graphed backpropagation diagrams of Figure 2.2. Each node is an entity, which locally stores all information relevant to its operation such as its output value (right side). This information was simply extended to also store the derivative (left side) and other attributes such as the sum of node activations and the string of the propagated analytical function. The link object stores bi-directional pointers to the two connected node objects to go forward (top step) and backward (bottom step). Each link objects also stores the weight and weight update (delta) value.

The nodes and links were both implemented as inherited classes. All nodes are derived from one base backpropagation node object (Figure A.1), where the basic characteristics like the error and activation function are defined. Each derived node (output, hidden, input, and bias) expands on or modifies the functionality inherited from the preceding node. Two link types are derived from the same base link class and only differ in their weight update rules to accommodate

APPENDIX A. TECHNICAL IMPLEMENTATION

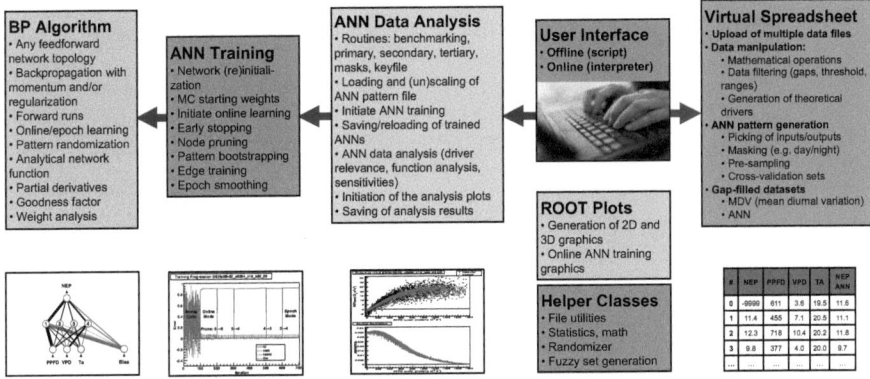

Figure A.2: Sketch of the five main modules (colored) plus two auxiliary modules (gray) of the program with a list of their core functions. (The diagrams below show a typical output.)

the two execution modes, online and epoch, of the backpropagation algorithm. The object-oriented network implementation is the heart of the modeling framework.

A.2 Program structure

The modeling framework has been embedded in a modular program with a hierarchical structure (Figure A.2). The top module is the User Interface, where commands can be entered offline in a script or directly online using the interpreter. The user commands access either the Virtual Spreadsheet or the ANN Data Analysis module.

The Virtual Spreadsheet module permits flexible pattern generation. Here, the response query is specified and driver candidates are generated, with a typical pattern generation script provided below (Section A.4).

The ANN Data Analysis module accommodates the automated processing of these generated patterns. It pipes via the ANN Training module to the BP Algorithm module, which contains the object-oriented network implementation described above. The following six processing routines for training and analyzing the ANN models were developed to implement the different phases of the methodology:

129

APPENDIX A. TECHNICAL IMPLEMENTATION

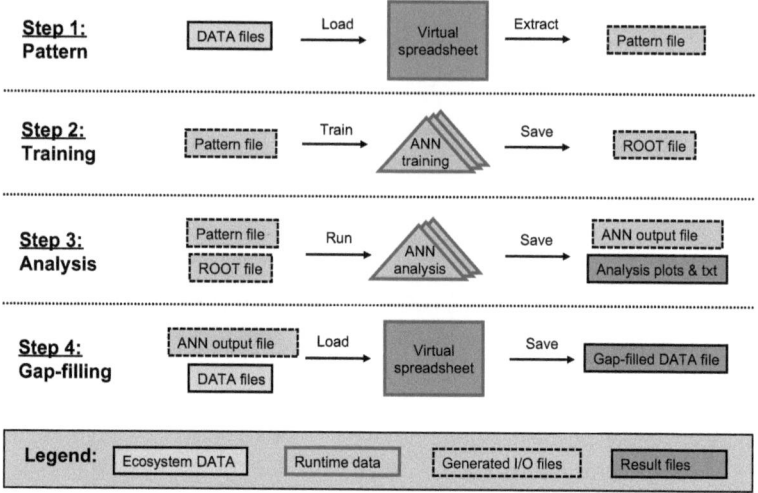

Figure A.3: Schematic of the four steps of the program.

b) Benchmark with all inputs,

p) Primary driver training on single inputs,

s) Secondary driver training on first plus secondary inputs,

t) Tertiary driver training on first two plus tertiary inputs,

m) Masking of data for grouping and binning, and

k) Keyfile scenario masks.

The labels in the beginning are used to specify the processing routine by the user. A typical specification ini to process a specific pattern in the ANN Data Analysis module is provided below (Section A.5).

The modules ROOT Plots and Helper Classes contain independent auxiliary functions that are commonly accessible.

A.3 Program flow

The flow of the program can be divided into the four steps presented in Figure A.3:

Pattern: Loading of the ecosystem data file(s) into the Virtual Spreadsheet. Generation of the pattern. Saving of the pattern dataset to a file.

Training: Loading of the pattern file into the ANN Data Analysis module. ANN trainings on the pattern (sub)sets. Saving of class instance of each trained ANN model and the training pattern information to a ROOT file.

Analysis: Reloading of the trained ANN models and pattern (sub)sets into the ANN Data Analysis module. Data analysis. Saving of the modeled ANN output and the analysis plots and results to files.

Gap-filling: For gap-filling, loading of the ANN output and the original ecosystem data files into the Virtual Spreadsheet. Generation of the gap-filled dataset. Saving of the gap-filled dataset to a file.

Each step can be executed and repeated separately. In Step 1, the pattern generation in the Virtual Spreadsheet is fully independent from the pattern processing.

Since the processing of the different routines in the ANN Data Analysis module produces multiple trained ANNs on specific pattern (sub)sets, it became necessary to also separate the training and analysis steps from each other. To perform for example the primary driver analysis shown in Figure 5.3, the ANNs are trained ten times on each of the twenty-five single drivers, resulting in a total of 250 individual trained ANN models. Therefore, in Step 2, the ANN model, its training properties, and its training pattern information are saved after each training process. This way, the time consuming training process can be interrupted and restarted any time. The trained ANN models are then (re)used for various setups of the analysis in Step 3.

In the last step, Step 4, the ANN model output can be used to gap-fill the original ecosystem dataset.

A.4 Example of a pattern generation script

To generate a pattern, direct user input or a script specifying the response query and the driver candidates is processed in the Virtual Spreadsheet. The queried response is extracted from the ecosystem dataset(s) by setting up filters, masks, ranges and thresholds. To generate the driver candidates, observed variables are picked or new theoretical variables can be computed. All variables that have been selected as input or output are written to a pattern file with a detailed

APPENDIX A. TECHNICAL IMPLEMENTATION

description of the pattern generation setting in its header. The data is saved in ascii format (*.txt) in a file named with a pattern-specific ID, consisting of the site name, the site years, and a unique pattern number.

As an example, the script to setup the pattern for characterizing the daytime NEP response of the Hainich forest in Chapter 5 is shown in Figure A.4. First, the ecosystem dataset is specified as a half-hourly dataset from the Carboeurope database with processing level 3 (CE3_h) of the Hainich site (DEHai) for the years 2000 to 2002 (00-02) and loaded (lines 2 and 3). For the canopy temperature, an auxiliary file is uploaded (line 4).

Next, the response query is specified. All but the best quality carbon flux data is filtered[1] (line 7), the column $NEP = -NEE$ added (line 8), and picked as an output (line 10). The data tuple range is limited[2] to the active season of the month June to September and to the daytime (half-hours with light) response (lines 11 to 16).

In the following lines, the driver candidates are generated. First all observed radiative and meteorological variables plus the provided potential radiation R_{pot} are picked as inputs (lines 20 and 21). Then the theoretical variables are generated, the diffuse fraction f_{dif} and diffuse and direct $PPFD$ (lines 23 to 27), the vapor pressure deficit VPD (line 28), the daily mean daytime temperature T_m (line 30), a fuzzy variable for the time of the day (line 31), and the past half-hour of NEP_{hh} (line 32). All of these are picked as inputs (line 34).

At the end of the script, the mask is set to flux bins (line 37) and the pattern is written to a file (line 38).

After running the script, the pattern used for the analysis in Chapter 5 was generated. The pattern had been assigned the pattern ID `DEHai00-02_p0264`, as can be seen in the header of the corresponding analysis plots.

A.5 Example of a typical specification file

To process a specific pattern, the user needs to specify the processing routine, the (subset of) inputs and outputs to use for the ANN data analysis, and the number of training repetitions. Each new setup of a processing routine and combination of inputs and outputs is assigned a new version number. For later

---
[1]Filtered data tuples will be saved to the pattern file (e.g. to be filled) but not used for ANN training.
[2]Data tuples within the limited range will be saved and the ones outside discarded.

APPENDIX A. TECHNICAL IMPLEMENTATION

```
1   //*** Upload the ecosystem data ***
2   Workbook wb("CE3_h","DEHai","00-02"); //set ecosystem dataset
3   wb.LoadMainFile(); //upload
4   wb.LoadAuxFile("f_Tc","DEHai","00-02"); //upload auxiliary dataset
5
6   //*** Specify the response query ***
7   wb.FilterData("qf_NEE_st",'>',0); //filter all but best quality (flag = 0)
8   wb.NegOfCol("NEE_st","NEP"); //add NEP column
9   wb.FilterData("NEP",'<',-10); //filter one extreme outlier
10  wb.PickCol('o',"NEP"); //pick NEP as output
11  wb.LimitRange("Month",'&',6,9); //limit to months June to Sept
12  wb.PickMultiCol('s',"YEAR Month"); //pick as standard (for reference)
13  wb.LimitRange("PPFD",'>',5); //limit to half-hours with light
14  wb.LimitRange("Rg",'>',1); //(just to make double sure)
15  wb.FilterData("PPFD",'\%',0,10); //filter up to 10 since often unclean
16  wb.LimitRange("qf_Rg",'=',0); //limit range to best quality (flag = 0)
17
18  //*** Generate driver candidates ***
19  //Pick the observed drivers plus the provided R_pot
20  wb.PickMultiCol('i',"SWC Rh Precip Ta Tc Ts1 Ts2 Gs ZL WD WS ustar");
21  wb.PickMultiCol('i',"PPFD Rg Rd Rr Rn R_pot"); //pick as inputs
22  //Generate the theoretical drivers
23  wb.FractionOf2Cols("Rd","Rg","f_dif"); //calculate f_dif
24  wb.LimitRange("f_dif",'{',1,GAP,ON); //only plausible values <1
25  wb.LimitRange("f_dif",'}',0.005); //discard unclean range around 0
26  wb.ProductOf2Cols("PPFD","f_dif","PPFDdif"); //calculate PPFDdif
27  wb.DifferenceOf2Cols("PPFD","PPFDdif","PPFDdir"); //calculate PPFDdir
28  wb.VPDCol("Ta","Rh"); //calculate VPD
29  wb.LimitRange("VPD",'<',18); //discard 5 outliers
30  wb.DailyMeanOfCol("Ta","T_m","PPFD",'d'); //calculate T_m
31  wb.Fuzzify('d',"Fuzzy"); //sinus curve for morning/afternoon
32  wb.PastHHOfCol("NEP","NEP_hh"); //add past half-hour to data tuples
33  wb.FilterData("NEP_hh",'<',-10); //filter again one outlier (see above)
34  wb.PickMultiCol('i',"f_dif PPFDdir PPFDdif VPD T_m Fuzzy NEP_hh");
35
36  //*** Set mask and write pattern to file ***
37  wb.SetMPC('f'); //set mask 'f' = masking of fluxes to bins of width 5
38  wb.WritePatternFile();
```

Figure A.4: Script to generate the pattern for characterizing the daytime NEP response of the Hainich forest in Chapter 5.

reference, the setup specifications are saved to a pattern ini-file

Figure A.5 shows the ini-file of pattern DEHai00-02_p0264 with the specifications, that were used to produce the various results presented in Chapter 5.

133

APPENDIX A. TECHNICAL IMPLEMENTATION

```
1  b  1  25  10  NEP  PPFD  PPFDdir  PPFDdif  Rg  Rd  Rr  Rn  Ta  T_m  Tc  Tsl ...
2  p  1  25  10  NEP  PPFD  PPFDdir  PPFDdif  Rg  Rd  Rr  Rn  Ta  T_m  Tc  Tsl ...
3  s  1  25  10  NEP  PPFD  PPFDdir  PPFDdif  Rg  Rd  Rr  Rn  Ta  T_m  Tc  Tsl ...
4  t  1  25  10  NEP  PPFD  PPFDdir  PPFDdif  Rg  Rd  Rr  Rn  Ta  T_m  Tc  Tsl ...
...
6  b  1   2  10  NEP  PPFDdir  PPFDdif
...
10 b  1   3  10  NEP  PPFDdir  PPFDdif  VPD
```

Figure A.5: `DEHai00-02_p0264.ini` with the specifications of the processing routines used to produce the various results presented in Chapter 5.

The first column is the setup version number, the second column the label of the processing routine described above, the third and fourth column are the number of outputs and inputs respectively, and the fifth column is the number of training permutations. The next columns are the variables names, starting with the outputs and followed by the inputs. The setup version #1 specifies for example the benchmarking (b) runs of Section 5.3 with 1 output, 25 inputs, and 10 training permutations. To (re)process setup version #1, the following two lines of code are executed:

    ANNanalysis eco_response; //initialize class
    eco_response.Process("DEHai","00−02",264,1);

The corresponding analysis plots and results are produced and saved automatically. The title of the plots and results files contains the type of plot, the pattern-specific ID, the version number of the training setup, the processing routine, and the training permutation. For example, the title of the scatterplot in Figure 5.1 generated by setup version v01 for the benchmarking run b00 and training permutation 00 was: `Output versus Target DEHai00-02_p0264_v01_b00_00`.

The technical implementation of the methodology resulted in a program with seven modules and over 15000 lines of code. The run time of the ANN data analysis is mainly determined by the training of the ANN models, more precisely by the size of the dataset (number of presented pattern tuples), the network structure (number of nodes and layers), and the training progression (number of iterations). The results presented in this manuscript were performed on a 2.4 Ghz Intel 2 dual core processor and one ANN training took from several seconds up to minutes.

Bibliography

Abramowitz G (2005) Towards a benchmark for land surface models. *Geophysical Research Letters*, **32**, L22702.

Abramowitz G, Pitman A, Gupta H, Kowalczyk E, Yingping W (2007) Systematic bias in land surface models. *Journal of Hydrometeorology*, **8**, 989–1001.

Atkin OK, Millar AH, Gardestrom P, Day DA, Leegood RC, Kennedy R (2000) Photosynthesis, carbohydrate metabolism and respiration in leaves of higher plants. In: *Photosynthesis: Physiology and Metabolism* (eds. Leegood RC, Sharkey TD, von Caemmerer S), Advances in Photosynthesis: Volume 9, pp. 153–175. Kluwer Academic Publishers, Dordrecht, The Netherlands.

Aubinet M, Chermanne B, Vandenhaute M, Longdoz B, Yernaux M, Laitat E (2001) Long term carbon dioxide exchange above a mixed forest in the Belgian Ardennes. *Agricultural and Forest Meteorology*, **108**, 293–315.

Aubinet M, Clement R, Elbers JA, *et al.* (2003) Methodology for data acquisition, storage, and treatment. In: *Fluxes of Carbon, Water and Energy of European Forests* (ed. Valentini R), Egological Studies 163, pp. 9–35. Springer-Verlag, Berlin Heidelberg.

Baldocchi DD (2008) Breathing of the terrestrial biosphere: lessons learned from a global network of carbon dioxide flux measurement systems. *Australian Journal of Botany*, **56**, 1–26.

Baly ECC (1935) The kinetics of photosynthesis. *Proceedings of the Royal Society of London Series B-Biological Sciences*, **117**, 218–239.

Bishop CM (1995) *Neural Networks for Pattern Recognition.* Oxford University Press, Oxford, United Kingdom.

BIBLIOGRAPHY

Blackman FF (1905) Optima and limiting factors. *Annals of Botany*, **19**, 281–296.

Brun R, Rademakers F (1997) ROOT - An object oriented data analysis framework. *Nuclear Instruments and Methods in Physics Research Section A*, **389**, 81–86. See also http://root.cern.ch/.

Canadell JG, Le Quere C, Raupach MR, et al. (2007) Contributions to accelerating atmospheric CO2 growth from economic activity, carbon intensity, and efficiency of natural sinks. *Proceedings of the National Academy of Sciences of the United States of America*, **104**, 18866–18870.

Cybenko GV (1989) Approximation by superpositions of a sigmoidal function. *Mathematics of Control, Signals, and Systems*, **2**, 303–314.

Desai AR, Bolstad PV, Cook BD, Davis KJ, Carey EV (2005) Comparing net ecosystem exchange of carbon dioxide between an old-growth and mature forest in the upper Midwest, USA. *Agricultural and Forest Meteorology*, **128**, 33–55.

Dimopoulos I, Chronopoulos J, Chronopoulou-Sereli A, Lek S (1999) Neural network models to study relationships between lead concentration in grasses and permanent urban descriptors in Athens city (Greece). *Ecological Modelling*, **120**, 157–165.

Falge E, Baldocchi D, Olson R, et al. (2001) Gap filling strategies for defensible annual sums of net ecosystem exchange. *Agricultural and Forest Meteorology*, **107**, 43–69.

Gevrey M, Dimopoulos L, Lek S (2003) Review and comparison of methods to study the contribution of variables in artificial neural network models. *Ecological Modelling*, **160**, 249–264.

Gilmanov TG, Verma SB, Sims PL, Meyers TP, Bradford JA, Burba GG, Suyker AE (2003) Gross primary production and light response parameters of four Southern Plains ecosystems estimated using long-term CO2-flux tower measurements. *Global Biogeochemical Cycles*, **17**, GB1071.

Goeckede M, Rebmann C, Foken T (2004) A combination of quality assessment tools for eddy covariance measurements with footprint modelling for the characterisation of complex sites. *Agricultural and Forest Meteorology*, **127**, 175–188.

Gove JH, Hollinger DY (2006) Application of a dual unscented Kalman filter for simultaneous state and parameter estimation in problems of surface-atmosphere exchange. *Journal of Geophysical Research-Atmospheres*, **111**, D08S07.

Gu LH, Baldocchi D, Verma SB, Black TA, Vesala T, Falge EM, Dowty PR (2002) Advantages of diffuse radiation for terrestrial ecosystem productivity. *Journal of Geophysical Research-Atmospheres*, **107**, D06–4050.

Hempel CG, Oppenheim P (1948) Studies in the logic of explanation. *Philosophy of Science*, **15**, 135–175.

Hollinger DY, Richardson AD (2005) Uncertainty in eddy covariance measurements and its application to physiological models. *Tree Physiology*, **25**, 873–885.

IPCC (2007) The Carbon Cycle and the Climate System (Chapter 7.3). In: *Climate Change 2007: The Physical Science Basis. Contribution of Working Group I to the Fourth Assessment Report of the Intergovernmental Panel on Climate Change*, pp. 511–539. Cambridge University Press, Cambridge, United Kingdom and New York, NY, USA.

Jassby AD, Platt T (1976) Mathematical formulation of relationship between photosynthesis and light for phytoplankton. *Limnology and Oceanography*, **21**, 540–547.

Knohl A, Baldocchi DD (2008) Effects of diffuse radiation on canopy gas exchange processes in a forest ecosystem. *Journal of Geophysical Research-Biogeosciences*, **113**, G02023.

Knohl A, Schulze ED, Kolle O, Buchmann N (2003) Large carbon uptake by an unmanaged 250-year-old deciduous forest in Central Germany. *Agricultural and Forest Meteorology*, **118**, 151–167.

Knorr W, Kattge J (2005) Inversion of terrestrial ecosystem model parameter values against eddy covariance measurements by Monte Carlo sampling. *Global Change Biology*, **11**, 1333–1351.

Kok B (1948) A critical consideration of the quantum yield of Chlorella-photosynthesis. *Enzymologia*, **13**, 1–56.

Krinner G, Viovy N, de Noblet-Ducoudre N, *et al.* (2005) A dynamic global vegetation model for studies of the coupled atmosphere-biosphere system. *Global Biogeochemical Cycles*, **19**, GB1015.

Kutsch WL, Kolle O, Rebmann C, Knohl A, Ziegler W, Schulze ED (2008) Advection and resulting CO2 exchange uncertainty in a tall forest in Central Germany. *Ecological Applications*, **18**, 1391–1405.

Larcher W (2003) *Physiological plant ecology*. Springer-Verlag, Berlin Heidelberg, 4th edn.

Lasslop G, Reichstein M, Kattge J, Papale D (2008) Influences of observation errors in eddy flux data on inverse model parameter estimation. *Biogeosciences*, **5**, 1311–1324.

Lasslop G, Reichstein M, Papale D, *et al.* (2009) Separation of net ecosystem exchange into assimilation and respiration using a light response curve approach: critical issues and global evaluation. *Global Change Biology*, **16**, 187–208.

Lek S, Guégan JF (2000) Use aspects (Chapter 1.2.6). In: *Artificial Neural Networks: Application to Ecology and Evolution*, pp. 12–13. Springer-Verlag, Berlin Heidelberg.

Levin I (2009) 50 years of atmospheric 14CO2 observations - a key for recent carbon cycle research. *Talk at ICDC 8, Jena*.

Looney CG (1997) Why Output Target Components Can be Neither 0 or 1 (Chapter 4.8.1). In: *Pattern Recognition Using Neural Networks*, p. 131. Oxford University Press, New York, NY, USA.

MacKay DJC (2003) Beyond descent on the error function: regularization (Chapter 39.4). In: *Information Theory, Inference and Learning Algorithms*, p. 479. Cambridge University Press, Cambridge, United Kingdom and New York, NY, USA.

Marti O, Braconnot P, Bellier J, *et al.* (2005) The new IPSL climate system model: IPSL-CM4. *Note du Pôle de Modélisation*, **26**, 84pp.

BIBLIOGRAPHY

Michaelis L, Menten ML (1913) Die Kinetik der Invertinwirkung. *Biochemische Zeitschrift*, **49**, 333–369.

Moffat AM (In preparation) Implications of modeling the daytime ecosystem CO2 response with one-dimensional light response curves.

Moffat AM, Papale D, Reichstein M, *et al.* (2007) Comprehensive comparison of gap-filling techniques for eddy covariance net carbon fluxes. *Agricultural and Forest Meteorology*, **147**, 209–232.

Murase K, Matsunaga Y, Nakade Y (1991) A backpropagation algorithm which automatically determines the number of association units. *IEEE International Joint Conference on Neural Networks*, pp. 783–788.

Ögren E (1993) Convexity of the photosynthetic light-response curve in relation to intensity and direction of light during growth. *Plant Physiology*, **101**, 1013–1019.

Papale D, Reichstein M, Aubinet M, *et al.* (2006) Towards a standardized processing of net ecosystem exchange measured with eddy covariance technique: algorithms and uncertainty estimation. *Biogeosciences*, **3**, 571–583.

Papale D, Valentini A (2003) A new assessment of European forests carbon exchanges by eddy fluxes and artificial neural network spatialization. *Global Change Biology*, **9**, 525–535.

Rabinowitch EI (1951) *Photosynthesis and related processes*, vol. II (1). Interscience Publishers, New York, NY, USA.

Raddatz TJ, Reick CH, Knorr W, *et al.* (2007) Will the tropical land biosphere dominate the climate-carbon cycle feedback during the twenty-first century? *Climate Dynamics*, **29**, 565–574.

Raupach MR, Rayner PJ, Barrett DJ, *et al.* (2005) Model-data synthesis in terrestrial carbon observation: methods, data requirements and data uncertainty specifications. *Global Change Biology*, **11**, 378–397.

Reed RD, Marks RJ (1999) Effects of Training with Noisy Inputs (Chapter 17). In: *Neural Smithing: Supervised Learning in Feedforward Artificial Neural Networks*, pp. 277–292. The MIT Press, Cambridge, MA, USA and London, United Kingdom.

Richardson A, Mahecha M, Falge E, et al. (2008) Statistical properties of random CO_2 flux measurement uncertainty inferred from model residuals. *Agricultural and Forest Meteorology*, **148**, 38–50.

Richardson AD, Hollinger DY (2007) A method to estimate the additional uncertainty in gap-filled NEE resulting from long gaps in the CO_2 flux record. *Agricultural and Forest Meteorology*, **147**, 199–208.

Richardson AD, Hollinger DY, Burba GG, et al. (2006) A multi-site analysis of random error in tower-based measurements of carbon and energy fluxes. *Agricultural and Forest Meteorology*, **136**, 1–18.

Ritchie R (2008) Fitting light saturation curves measured using modulated fluorometry. *Photosynthesis Research*, **96**, 201–215.

Rogers J (1996) *Object-oriented neural networks in C++*. Academic Press, San Diego, CA, USA.

Rojas R (1996) *Neural Networks - A Systematic Introduction*. Springer-Verlag, Berlin Heidelberg.

Rumelhart DE, Hinton GE, Williams RJ (1986) Learning representations by back-propagating errors. *Nature*, **323**, 533–536.

Schmid HP (1994) Source areas for scalars and scalar fluxes. *Boundary-Layer Meteorology*, **67**, 293–318.

Schulze ED (1970) Der CO2-Gaswechsel der Buche (Fagus silvatica L.) in Abhängigkeit von den Klimafaktoren im Freiland. *Flora*, **159**, 177–232.

Smith EL (1937) The influence of light and carbon dioxide on photosynthesis. *Journal of General Physiology*, **20**, 807–830.

Smith EL (1938) Limiting factors in photosynthesis: light and carbon dioxide. *Journal of General Physiology*, **22**, 21–35.

Stauch VJ, Jarvis AJ (2006) A semi-parametric gap-filling model for eddy covariance CO_2 flux time series data. *Global Change Biology*, **12**, 1707–1716.

van de Laar P, Heskes T, Gielen S (1999) Partial retraining: a new approach to input relevance determination. *International Journal of Neural Systems*, **9**, 75–85.

Wang YP, Trudinger CM, Enting IG (2009) A review of applications of model-data fusion to studies of terrestrial carbon fluxes at different scales. *Agricultural and Forest Meteorology*, **149**, 1829–1842.

Wolkoff A (1866) *Ueber die Einwirkung des Lichtes auf Pflanzen*. Magisterarbeit, Physico-mathematische Facultät, Kaiserliche Universität zu Dorpat.

Young PC, Jarvis AJ (2002) Data-based mechanistic modelling and state dependent parameter models. *CRES Report*, **TR/177**, 37pp.

Zadeh LA (1965) Fuzzy Sets. *Information and Control*, **8**, 338–353.

Zaehle S, Friend AD (2010) Carbon and nitrogen cycle dynamics in the O-CN land surface model: 1. Model description, site-scale evaluation, and sensitivity to parameter estimates. *Global Biogeochemical Cycles*, **24**, GB1005.

Zhang LM, Yu GR, Sun XM, *et al.* (2006) Seasonal variations of ecosystem apparent quantum yield (alpha) and maximum photosynthesis rate (P-max) of different forest ecosystems in China. *Agricultural and Forest Meteorology*, **137**, 176–187.

i want morebooks!

Buy your books fast and straightforward online - at one of world's fastest growing online book stores! Environmentally sound due to Print-on-Demand technologies.

Buy your books online at
www.get-morebooks.com

Kaufen Sie Ihre Bücher schnell und unkompliziert online – auf einer der am schnellsten wachsenden Buchhandelsplattformen weltweit! Dank Print-On-Demand umwelt- und ressourcenschonend produziert.

Bücher schneller online kaufen
www.morebooks.de

VDM Verlagsservicegesellschaft mbH
Heinrich-Böcking-Str. 6-8
D - 66121 Saarbrücken

Telefon: +49 681 3720 174
Telefax: +49 681 3720 1749

info@vdm-vsg.de
www.vdm-vsg.de

Printed by Books on Demand GmbH, Norderstedt / Germany